WindEurope Electric City 2021

Journal of Physics: Conference Series Volume 2151

Copenhagen, Denmark
23 - 25 November 2021

Editor:

Michael Muskulus

ISBN: 978-1-7138-5122-6
ISSN: 1742-6588

Printed from e-media with permission by:

Curran Associates, Inc.
57 Morehouse Lane
Red Hook, NY 12571

Institute of Physics
Dirac House, Temple Back
Bristol BS1 6BE UK

Phone: 44 1 17 929 7481
Fax: 44 1 17 920 0979

techtracking@iop.org

Additional copies of this publication are available from:

Curran Associates, Inc.
57 Morehouse Lane
Red Hook, NY 12571 USA
Phone: 845-758-0400
Fax: 845-758-2633
Email: curran@proceedings.com
Web: www.proceedings.com

TABLE OF CONTENTS

PREFACE

This Special Issue presents selected contributions from the **WindEurope Electric City 2021 Conference** that took place 22-26 November in Copenhagen. A total of 14 papers are included, addressing one (or more) of the three conference themes:

- Onshore wind
- Offshore wind
- Electrifying the energy system.

Presenters in the Science & Research Track of the conference were invited to submit a paper for these proceedings. All submissions have gone through a careful peer-review organized by the European Academy of Wind Energy (EAWE[1]).

Conference committee for the scientific programme

- Stephan Barth, Managing Director, ForWind - Center for Wind Energy Research
- Ola Carlson, Professor, Chalmers University
- James Carroll, Senior Lecturer, Strathclyde University
- Georges Kariniotakis, Professor, MINES ParisTech
- Athanasios Kolios, Professor, Strathclyde University
- Jakob Mann, Professor, DTU Wind Energy
- Michael Muskulus, Professor, Norwegian University of Science and Technology NTNU
- Simon Watson, Professor, Delft University of Technology

We thank all 20+ reviewers for their engagement and effort, without which these proceedings would not have been possible.

As with previous WindEurope events, the European Academy of Wind Energy has been responsible for organizing the review process for scientific contributions and has contributed to the development of the conference sessions. WindEurope has been responsible for the organization and logistics of the conference, and we thank their highly professional staff and their associates for the excellent collaboration.

Proceedings editor
Michael Muskulus
Professor, Norwegian University of Science and Technology NTNU

[1] www.eawe.eu

Peer Review Statement

All papers published in this volume have been reviewed through processes administered by the Editors. Reviews were conducted by expert referees to the professional and scientific standards expected of a proceedings journal published by IOP Publishing.

- **Type of peer review:** Single Anonymous
- **Conference submission management system:** Morressier
- **Number of submissions received:** 15
- **Number of submissions sent for review:** 15
- **Number of submissions accepted:** 14
- **Acceptance Rate (Submissions Accepted / Submissions Received × 100):** 93.3
- **Average number of reviews per paper:** 3
- **Total number of reviewers involved:** 20
- **Contact person for queries:**
  Name: Michael Muskulus
  Email: michael.muskulus@ntnu.no
  Affiliation: Norwegian University of Science and Technology NTNU

Journal of Physics: Conference Series **2151** (2022) 012001 doi:10.1088/1742-6596/2151/1/012001

Induced stalled flow due to roughness sensitivity for thick airfoils in modern wind turbines

R Gutiérrez[1], **E Llorente**[1] **and D Ragni**[2]

[1]Nordex Energy Spain S.A.U, Avenida Ciudad de la Innovación 3, 31621 Sarriguren, Navarra, Spain.

[2]Aerodynamics, Wind Energy, Flight Performance and Propulsion Department, Faculty of Aerospace Engineering, Delft University of Technology, Kluyverweg 1, 2629, HS Delft, the Netherlands.

E-mail: `RGutierrez@nordex-online.com`

Abstract. The mid-span region of wind turbine blades can be thickened to fulfil the structural requirements of the blade. Hence, thick airfoils, that were designed to operate at the root region of the blade, are moved to the mid-span region. This could not imply remarkable variations of the blade performance once its surface is smooth. However, the sensitivity of thick airfoils to roughness could cause significant aerodynamic impacts such as flow separation. This research aims to quantify the impact of the blade thickness, under smooth and rough conditions, in the annual energy production and the fatigue loads of the blade. Ten blade designs, linearly interpolated in thickness, are studied employing aero-elastic computations. The results reveal that the thickest blade increases the annual energy production by 5% with respect to the thinnest blade under rough conditions. Whereas this increase is less than 1% under smooth conditions. The loss of annual energy production varies with the blade thickness linearly for thin blades while it varies exponentially for thick blades up to 22%. Fatigue loads assessment confirmed a reduction of the damage equivalent load under smooth conditions, whereas the thickest blade increased it 28% under rough conditions.

1. Introduction.

The surface roughness of wind turbine blades concerns wind turbine manufacturers from the characterization of the roughness on the blade [1] to the estimation of the wind turbine aerodynamic performance [2]. In parallel, the constant strive to extract more energy from the wind have encouraged blade designers to increase the length of wind turbine blades [3]. As explained by Jensen et al. [4], the weight of the blades should increase with blade length to the power of three. As a result, the effect of gravity and inertial loads gets dominant with respect to aerodynamic loading. Additionally, this increases the blade stiffness requirements necessary to withstand loads and keep tip deflection within limits.

One design approach for achieving the same level of stiffness with less material is to increase the thickness of the blade. The displacement of the spar cap from the neutral axis enhances the blade structural properties. However, this approach moves thick airfoils from the inboard blade region to the outboard one. Whereas the performance of smooth blades can be slightly affected, a significant impact can occur for rough ones. As explained by Timmer et al. [5], thick airfoils can provide with premature flow separation in case of leading-edge soiling. Consequently, roughness sensitivity increases when thicker airfoils are required for the outboard sections of the blade.

For thin airfoils applications, roughness is known to decrease the maximum lift coefficient (C_{Lmax}), the slope of the lift coefficient C_L and increase the drag coefficient C_D [6]. Thick airfoils also suffer these aerodynamic impacts at high angles of attack [7]. However, Gutierrez et al. [8] measured an abrupt increase in C_L and C_D at low angles of attack which was caused by flow separation on the pressure side of a thick airfoil. This flow separation at low angles of attack was not previously reported in the literature because thick airfoils were designed to be insensitive to roughness at the high angles of attack of the root region. The use of thick airfoils in the mid-span region of the blade lowers the angle of attack (AoA) at which these airfoils operate. Consequently, the occurrences of flow separation on the blade pressure side can increase and influence the wind turbine performance. On the other hand, most of the studies about roughness impact on wind turbine performance, do not contemplate this kind of flow separation as they are mainly based on reporting Annual Energy Production (AEP) losses. Bak et al. [9] recently estimated an AEP loss of 5% for a 10 MW pitch controlled wind turbine.

This study aims to assess the influence of the blade thickness on the blade sensitivity to roughness.

2. Methods.

2.1. Description of sensitivity study.

This research carries out a sensitivity study to quantify the influence of the blade thickness on power production and fatigue loads of the blade for clean and rough conditions. For rough conditions some studies such as [28] and [29] assume the last 30% of the blade length to be eroded, this study assumes the last 50% to be rough. This decision is made to include the mid-span region of the blade in consideration. Even though, this could give a worst-case scenario in an erosion situation, other sources of roughness such as dirtiness could act at the mid-span region of the blade.

The relevant steps of this sensitivity study are shown in Figure 1.

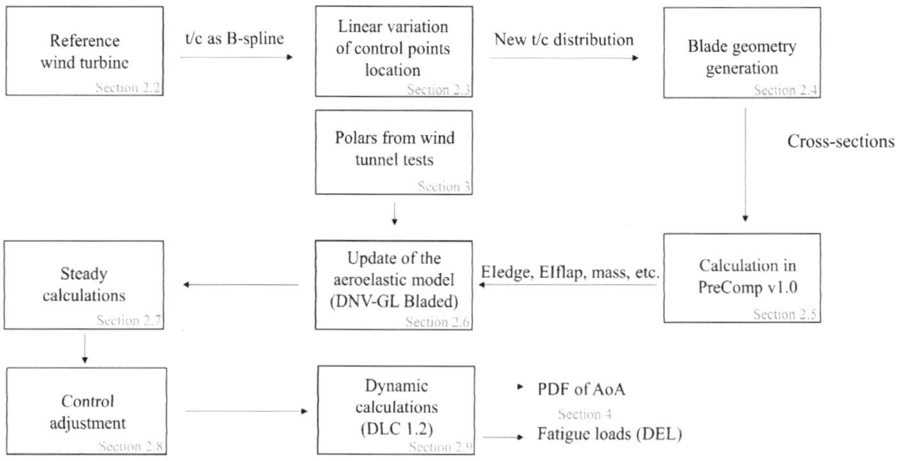

Figure 1: Sensitivity study procedure. The steps are related to the sections of this manuscript.

The variation of the t/c along a direction normal to the blade root (z or spanwise direction) is expressed as a B-spline function. The location of the control points of the B-spline is varied to obtain a different t/c distribution. As a next step, the blade geometry is expressed as a B-spline surface. Several cross sections are extracted from this surface as a list of points. These sections are used in the PreComp v1.0 software to compute their structural properties. At this last step, the number of laminate layers is equal to the one of the reference wind turbine. This reference wind turbine is defined in section 2.2. The relative location of the internal blade structure remains the same and its elements are adapted to the new blade geometry. These two last simplifications ensure that the variations in the results only depend on the blade thickness. The description of the blade geometry and the cross-sectional structural properties of the blade are updated in the aeroelastic model defined in the DNV-GL Bladed software [17]. As a result of this process, aeroelastic steady calculations are carried out for a total amount of ten blades. Later, the gain K_λ and minimum pitch angle β_{pitch} are adapted to reduce the impact on AEP for these blades. This gain refers to the relation between the generator torque and the square of the angular velocity of the rotor, as shown in Equation 2. After assessing the tendency of the impact on AEP, three blades are selected as the most representative ones. Finally, dynamic calculations are performed for these blades providing results of fatigue loads as well as AoA probability density function (PDF).

The steps shown in Figure 1 are explained in the following sections.

2.2. Reference wind turbine

A pitch controlled multi-megawatt wind turbine is used in this study. The relevant parameters of the wind turbine are summarized in Table 1.

Quantity	**Value**	**Units**
Blade length (L)	75.8	m
P_{rated}	4.5	MW
Control type	Pitch feathering	
Hub height	120	m
Rotor diameter	155.18	m
Rotor Cone	5	deg
Rotor Tilt	5	deg
Transmission	Gearbox	
Rotor position	Upwind	
Cut-in	3	m/s
Cut-out	25	m/s
IEC Class	S	

Table 1: Wind turbine model description.

2.3. Linear variation of control points location.

This study takes the advantage of the affine invariance feature of B-splines [18]. The t/c distribution of the reference blade is expressed as a 1D parameterized curve in B-spline. The ordinates of the control points of this B-spline are contained in the vector Q. The values of Q are varied to define a different distribution. This variation is defined as a linear interpolation between two bounds (Equation 1). These bounds delimit the design space and were built from the t/c distribution of the reference wind turbine.

$$Q(F) = (1 - F) \cdot Q_{low} + F \cdot Q_{up} \tag{1}$$

where F is the interpolation factor defined between 0 and 1. This factor determines the similarity of the t/c distribution with the upper and lower bounds distributions. A null value of F implies a t/c distribution equal to the lower bound. Otherwise, a unit value of F corresponds to the upper bound. Q_{up} and Q_{low} are the ordinates of the upper and lower bounds, respectively.

These bounds have been defined according to:

- t/c values are restricted to be $\leq$ 30% for z/L $\geq$ 0.5. This decision is based on the lack of empirical data in the rough case for airfoils with t/c higher than 30%.
- The first fourth control points are not modified to conserve the same blade root diameter for all blades.

The F factor was varied to obtain ten blades used in steady calculations. Three of these blades were selected for dynamic calculations whose t/c distribution is shown in Figure 2a. Reference wind turbine blades [10],[11],[12] are inside the order of the t/c values of the design space for the outer half of the blade length.

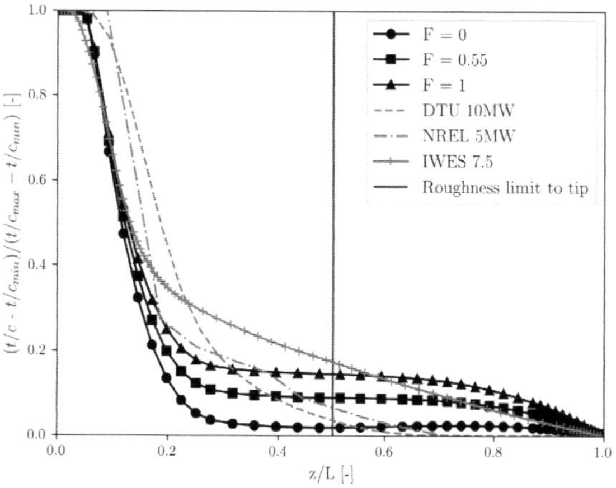

(a) Normalized relative thickness.

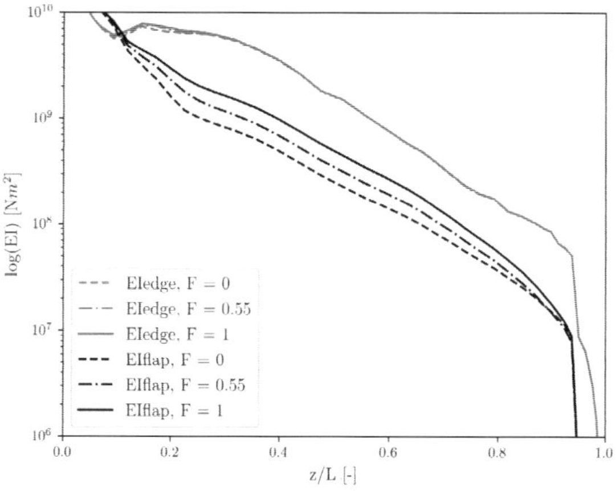

(b) Flapwise and edgewise stiffness.

Figure 2: Influence of the blade relative thickness on the stiffness of three blades of the study.

2.4. Blade geometry generation.

In this study, a set of predefined airfoils is transformed with respect to longitudinal distributions of the blade geometry variables. These distributions are expressed in B-spline form and determine the value along z of the t/c, the chord, the twist, the pre-bending and the relative airfoil location to z axis. This study refers to the predefined set of airfoils as master airfoils which are also treated in B-spline form. The t/c law dictates the location of these master airfoils and the cross-sectional variation between them. The t/c value at a precise z location is used as an interpolation factor for the cross-sectional variation between two consecutive master airfoils. Finally, the blade surface is expressed as a lofted B-spline surface [13].

2.5. Calculation in PreComp v1.0.

Cross-sectional structural properties are needed for the 1D Beam model used in Bladed [17]. PreComp v1.0 from NREL is the software chosen to perform the calculations of these properties. The code includes an analysis option to obtain blade properties such as bending stiffness, torsion stiffness, mass, moments of inertia, elastic–axis offset, and centre of mass offset along the blade

[16]. A two shear web configuration is used for the blade structure lay up of this study. The blade is assumed to be composed of different glass-fibre fabrics along with core materials infused in a matrix of epoxy resin. To study only the influence of the blade geometry on the results, the number of laminate layers per blade section is kept equal to the one used in the reference wind turbine.

Figure 2b compares the bending stiffness in flapwise and edgewise directions from PreComp calculations. According to [16], the EI_{edge} and EI_{flap} stiffness, are computed about a direction normal and parallel to the chord line of each section, respectively. The higher the F value is, the thicker the cross section is. Consequently, the laminate is relatively more displaced from the neutral axis and the stiffness is enhanced. This effect is notable from z/L values of 0.06 to 0.8. A higher increase of stiffness is given in the flapwise direction than in the edgewise direction.

2.6. Update of the aeroelastic model.

The new blade geometry description as well as the cross-sectional structural properties are specified in the aeroelastic model for a total number of 50 cross-sections. The polar data is specified only for the locations of the master airfoils. In this study, the aerodynamic coefficients data for intermediate values of t/c is interpolated with respect to the t/c value of two adjacent master airfoils.

For the rough blade cases, the specification of the aerodynamic data interpolation is adapted to fulfil the assumed roughness extension along the spanwise direction. As a result, clean and rough data are exchanged between the master airfoils depending on the t/c law. In Section 3, the empirical methodology, employed to extract this data, is explained.

For this study, the aerodynamic coefficients of the blade sections are determined for a Reynolds number (Re) of 3 million. This Re value corresponds to the maximum allowable value for the airfoil chord length and the wind tunnel explained in Section 3. The empirical data should be extrapolated in case of introducing the effect of Re variation with radial location and wind speed. However, there are no clear extrapolations to account for Re effect on the aerodynamic coefficients, especially for the rough conditions. The use of a wrong extrapolation would introduce more uncertainty into the study. A constant Re of 3 million along the blade, will provide with a worst-case scenario. But, variations between smooth and rough conditions are still valid because this Re assumption is used for both conditions.

2.7. Steady calculations.

Steady calculations are performed for uniform steady winds. Additionally, the deflections of the flexible blade components do not vary in time.

The result from these calculations is used to obtain the AEP of the ten studied blades. For the AEP calculation, a Weibull distribution is used with a shape factor k of 2 and an average wind speed at the hub height of 7.2 m/s.

2.8. Control adjustment.

The geometrical modification of the studied blades implies notable differences in structural properties such as a variation of the torsional stiffness and torsional flexibility. Hence, the range of operational angles of attack may differ and provide with power production losses with respect to the optimum performance. To overcome this impact, two basic control parameters, the gain K_λ and the minimum pitch angle β_{min}, are adjusted to ensure the maximum energy production per studied blade.

In regard to the Bladed theory manual [17], the demanded torque below rated power could be defined as:

$$T_g = K_\lambda \omega^2 \tag{2}$$

where K_λ is the gain and ω is the measured generator speed. This torque will maintain a desired tip-speed ratio λ and power production coefficient $C_p(\lambda)$.

Additionally, the power production below rated power depends on the minimum pitch variable β_{min}. This variable is a constant pitch value set for velocities lower than rated velocity. Above rated velocity, the pitch value is varied with respect to the pitch law.

Values of K_λ and β_{min} of 0.7 and 0 respectively are taken as a reference in clean calculations. However, an adjustement is done for rough cases to find the K_λ and β_{min} values that give the maximum AEP per blade design.

2.9. Dynamic calculations (DLC 1.2).

This study analyses the roughness impact on loads in terms of fatigue. Among the load cases described in IEC 61400-1 standard [15], DLC 1.2 cases are selected because they embody the requirements for loads that occur during the normal operation of a wind turbine throughout its lifetime. Specific situations such as start-up, shutdown and parked position are out of the scope of this research.

DLC 1.2 cases are defined by a normal turbulence model [15]. The wind speed standard deviation σ_1 is computed using the 90% quantile for a given hub height wind speed (V_{hub}) as suggested by the guideline. Consequently, for each wind speed, the turbulence intensity (TI) level is defined as σ_1/V_{hub}. Wind speed varies from 4 m/s to 24 m/s in steps of 2 m/s. Additionally, three yaw alignments, $-8°$, $0°$ and $8°$ are taken into account. A total number of 66 dynamic cases are computed with a simulation time per case of 10 minutes.

These cases are used to generate a stress history necessary to compute the cumulative damage of the wind turbine along its lifetime. The rainflow cycle counting subroutine of Bladed [17] is used to post-process the stress history. Finally, the computation of the damage equivalent load (DEL) is carried out for a wind turbine lifetime of 20 years.

3. Experimental database.

Available wind tunnel measurements focused on distributed roughness on airfoil leading edges (see Figure 3), are gathered to obtain the polar curves used in aero-elastic calculations for airfoils with t/c of 30%, 25% and 18%. The Reynolds number is 3 million for all the used experiments.

These aerodynamic measurements were carried out in the Low-Speed Lab Low-Turbulence Wind Tunnel (LSL-LTT) of the Delft University of Technology. LSL-LTT is an atmospheric tunnel of the closed-throat single-return type, with a contraction ratio of 17.8. It has a 2.9 m diameter six-bladed fan driven by a 525 kW DC motor, giving a maximum test section velocity of about 120 m/s. For all tests, a free stream velocity of 75 m/s was kept constant with a turbulence level of 0.07%. The test section is 1.80 m wide, 1.25 m high and 2.60 m long. Electrically actuated turntables flush with the test-section top and bottom walls positioned and supported a two-dimensional model in the middle of the test section length. All models were 1.25 meters long in the spanwise direction and 0.6 meters long in the chordwise direction. They were manufactured in resin infused glass fibre.

The angle of attack range is defined between $-20°$ and $20°$. A total number of 90 pressure taps were used on the models, to integrate the pressure component of the aerodynamic coefficients during 10 seconds of measuring time. The pressure taps orifices were at mid-span location and inclined to the incoming flow $15°$ to avoid the upstream influence of adjacent pressure taps. These orifices were more distributed on the airfoil leading-edge to retrieve the pressure gradient. A wake-rake of pitot tubes was installed downstream the airfoil to obtain total C_D based on the loss momentum method [19]. However, for angles of attack with flow separation, only pressure drag component C_{Dp} from pressure taps is used.

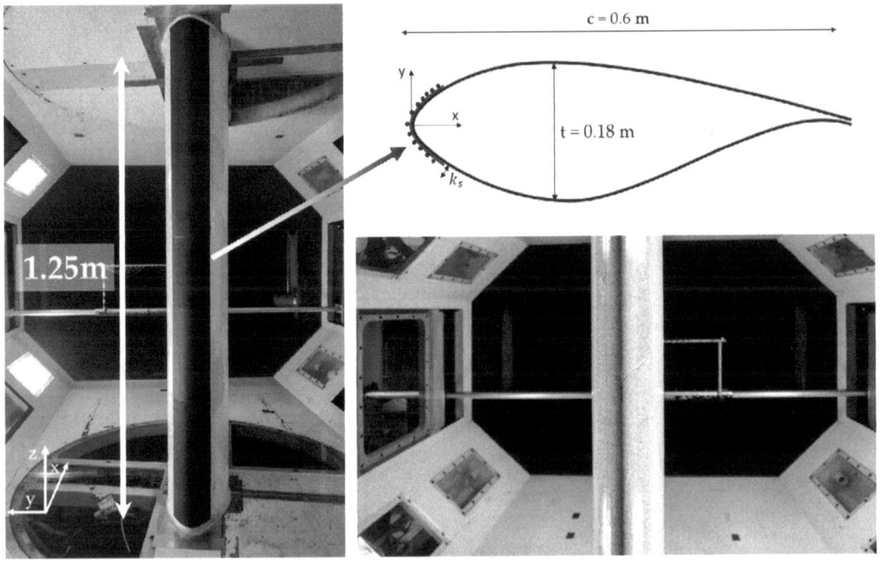

Figure 3: Sandpaper like roughness on the left and its drawing on top. Blown carborundum on the bottom right side.

3.1. Roughness emulation 25% and 18% thick airfoils.

To emulate the roughness state on the airfoil leading-edge, carborundum grains were blown on a double-sided tape. This tape was attached to the first 8%c on both airfoil sides following NACA standard method [6]. The grains density was ensured to be between 10% and 15% of the covered area. The normalized average grain height k_s/c was $4.67 \cdot 10^{-4}$ for both airfoils.

3.2. Roughness emulation 30% thick airfoil.

Sandpaper was used to emulate the roughness state on the airfoil leading-edge similarly to the work of Pires et al. [20]. The sandpaper was attached to the first 8%c on both airfoil sides. This sandpaper corresponded to an normalized average grain size (k_s/c) of $1.16 \cdot 10^{-3}$.

The main difference of this method with the 18% and 25% airfoil is the average grain height and density. Besides, sandpaper provides with a higher density of grains than the method of blown carborundum. On the other hand, another value of grain size is used.

It is important to remark that for all roughness states, no correlation is used in this study between the emulated roughness height and a roughness height found in a real wind turbine. The scientific community is researching this correlation [25], [23], [24]. The relatively higher value of k_s/c of the 30% thick airfoil, is combined with larger chords at the mid-span region of the blade. This would result in a higher k_s value than at outboards sections. As described by Wilcox et al. [26], thick airfoils collect larger regions of insect impingement due to the relatively lower impact velocity. In addition, Krisnhan et al. [27] demonstrated that insects which impact with high angles are less likely to smear over the surface. As a result, a higher height of attached particles is expected on the thick airfoils surface which is in concordance with the k_s/c value.

4. Results.

4.1. Empirical results.

The comparison of the variation of C_L and C_D, shown in Figure 4, shows that the thickest airfoil provides with the most significant C_D increase and C_L loss for positives angles of attack once the leading-edge gets rough. For angles of attack below 5°, the C_L and the C_D of the 30% airfoil significantly increases compared to the other airfoils. For these low angles of attack the thick airfoil geometry generates an adverse pressure gradient that separates the flow on the airfoil pressure side under rough conditions. Figure 5 shows this flow separation for both an infrared thermography (IR) image and C_p measurements. The IR image corroborates the location of the flow separation. The cause of the increase in C_L and C_D is the plain C_p distribution measured from the 50% of the airfoil chord length to the trailing-edge. The consequences of the difference in aerodynamic coefficients are analyzed in Section 4.3.

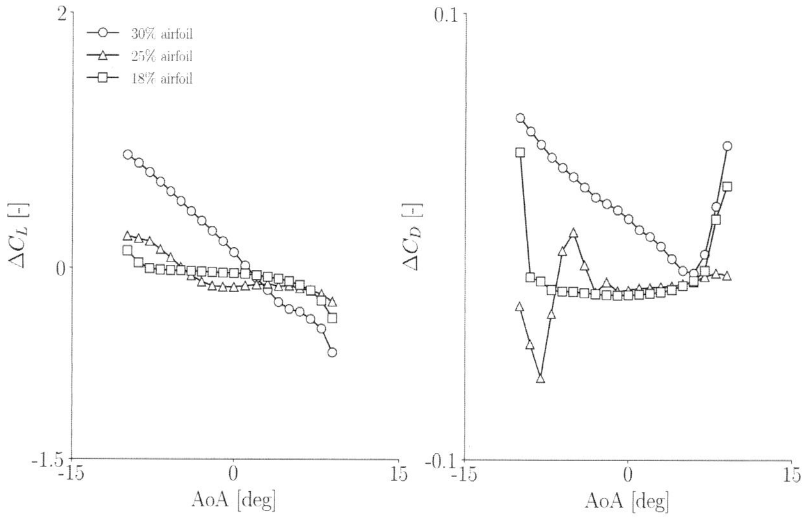

Figure 4: C_L loss and C_D variation with respect to smooth surface.

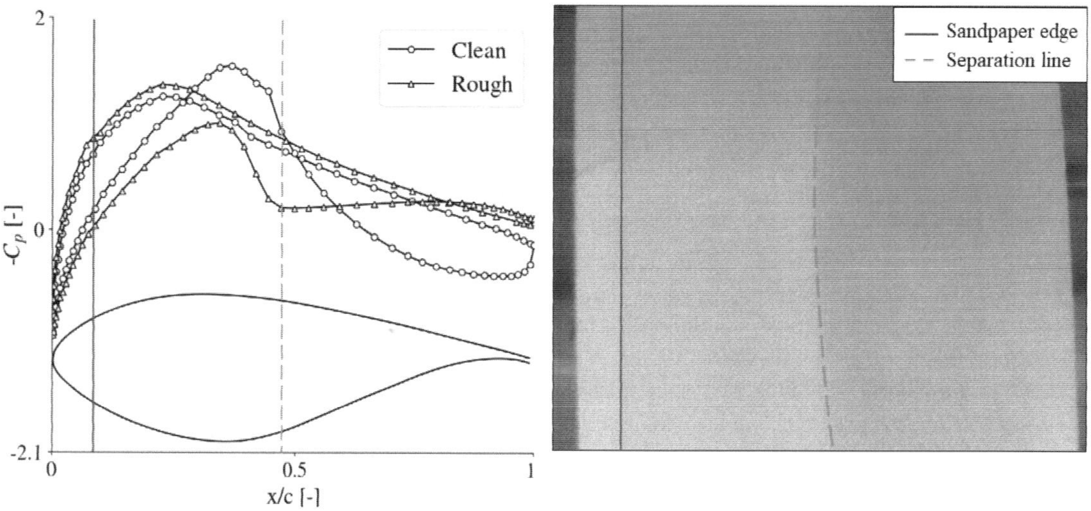

(a) Clean and rough C_p distributions.

(b) IR image on the rough airfoil pressure side.

Figure 5: Cross-check of the flow separation on the 30% thick airfoil at AoA of 0°.

4.2. Steady performance.
Figure 6 shows two different comparisons of the AEP. Figure 6a explains the impact of the increase in blade thickness for a given blade surface state (ΔAEP). This variable is computed with respect to the AEP of the thinnest blade (F = 0). The ΔAEP is at a maximum of 0.2 % in the smooth surface case, whereas, it abruptly changes in the rough case until a maximum value of 5%. The rough aerodynamic data of the 30% thick airfoil is used once the t/c value is 30% at the second half of the blade length. Blades with $F \geq 0.55$ fulfill this condition. This explains the abrupt change in the tendency from 1% to 5%. Consequently, the increase in blade thickness slightly varies the AEP under smooth surface conditions, whereas the AEP significantly varies under rough surface conditions.

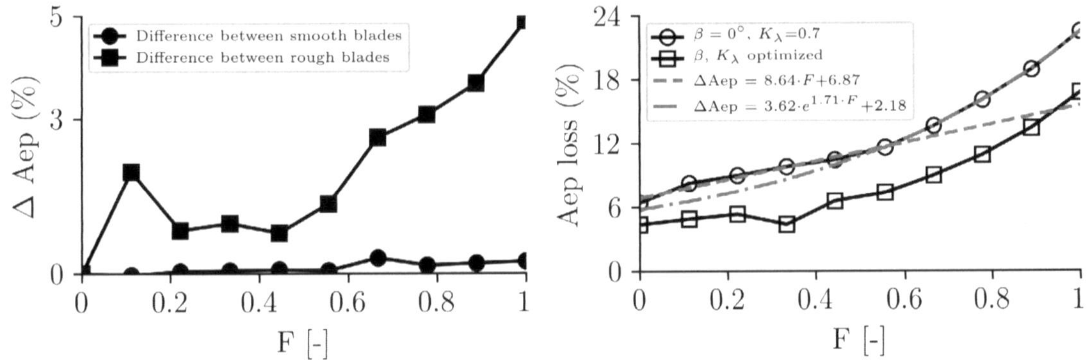

(a) (b)

Figure 6: The AEP variation with respect to the thinnest blade of the study is shown in (a) per each surface condition. The AEP loss of each blade with and without control optimization is shown in (b).

Besides, Figure 6b shows the AEP loss per blade design with respect to smooth surface conditions. A linear tendency of the AEP loss with respect to blade thickness increase is shown until an F value equal to 0.55 is reached. From this value, the AEP loss changes to exponential tendency reaching values until 22% in the thickest blade design (F=1). Additionally, the optimization of K_λ and β_{min} explained in Section 2.8 reduces the AEP loss as an average of 5%.

4.3. Separated flow consequences.
In section 4.1, flow separation occurs at angles of attack $\leq 5°$ resulting in abrupt changes of the aerodynamic coefficients which can provide with an impact on blade operation. Hence, the angle of attack determines the necessary conditions for flow separation. Additionally, this flow condition only takes place for the 30% thick airfoil under rough conditions. It only occurs for blade designs in which the 30% thick airfoil relies on the outer half of the blade. This occurs for F values $\geq$ 0.55.

Due to the relation of the angle of attack with flow separation, the probability density function of the angle of attack is computed for different blade sections. Dynamic calculations are used to extract PDFs. This is done for three out of ten blades of the study. These blades are the thinnest blade (F = 0), an intermediate blade (F = 0.55) and the thickest blade (F = 1). Their t/c law can be seen in Figure 2. Figure 7 shows a higher variation of the maximum peak of the PDF for the thickest blade. The z/L section of value 0.43 is inside the smooth blade region and the other two are inside the rough one. The maximum peak is moved to the left in the rough case with respect to the clean case. The likelihood of having separated flow is increased from F values higher than 0.55 as the maximum peak is around 5°. In contrast, the thinnest blade shows minimal variation in PDF distribution between surface states.

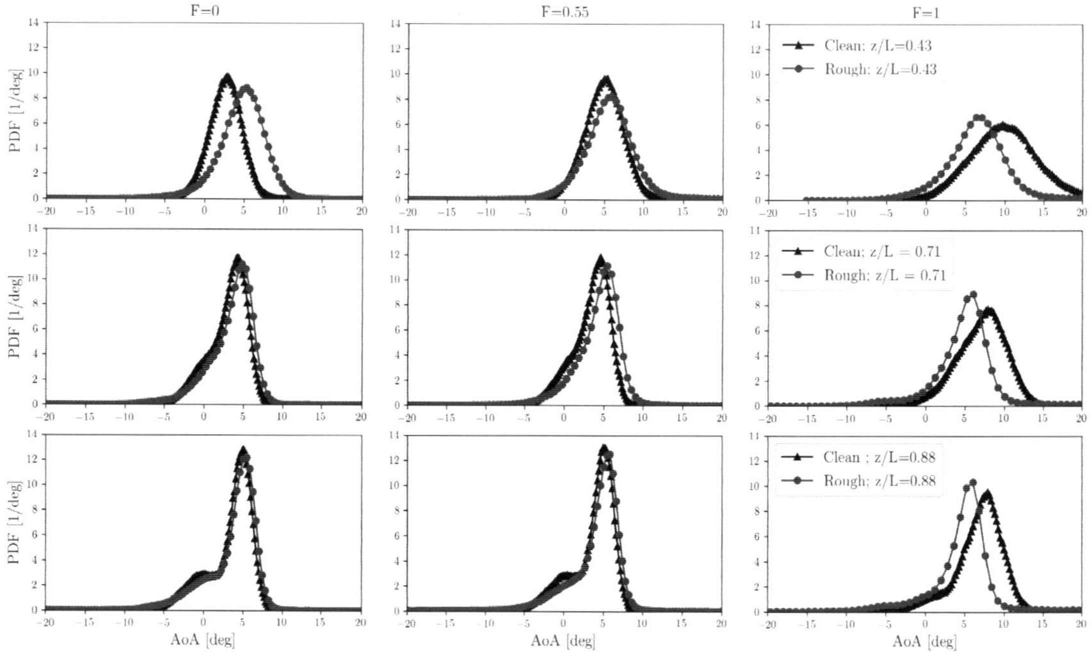

Figure 7: Probabilistic density functions (PDFs) for different relative blade locations z/L and F numbers.

After evaluating the probability of flow separation occurrence, its consequence is quantified by means of DELs. These loads were computed for three selected blade designs using the subroutine of Bladed software and the set-up described in section 2.9. As remarked by Papi et al. [28], the out-of-plane blade root bending moment M_{yroot} is a suitable indicator of overall blade loading. Additionally, the force F_{xroot} is relevant for the dimensioning of the root bearing. The x and y directions are defined normal and parallel to the chord line of each blade section, respectively. For the root section, x is parallel to the nacelle direction and y is normal to it. DELs are computed for M_{yroot} and F_{xroot} as these loads are related to the variation of EI_{flap} stiffness shown in Figure 2.

Figure 8 shows that the DEL of F_{xroot} is monotonically reduced in the clean case due to the effect of the blade thickness. This is related to the increase in blade stiffness shown in Figure 2. In contrast, the reduction of M_{yroot} is not monotonic but the thickest blade still provides with the lowest DEL value. For all blades, the rough surface state increases the DEL value with respect to clean blade condition. The thickest blade provides with the highest increment, 28% in F_{xroot} and 40% in M_{yroot}.

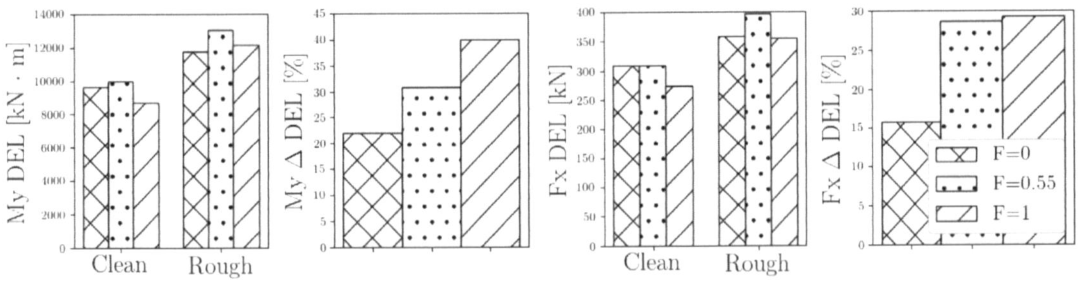

Figure 8: DEL comparison for 3 out of 10 blades of the studied. ΔDEL is computed with respect to clean value.

5. Conclusions.

This study quantifies these consequences in terms of AEP, DELs and likelihood of having flow separation in the case of thick blades.

Experimental data reveals that the 30% thick airfoil provides with notable sensitivity to roughness compared to 18% and 25% thick airfoils. Flow separation on the airfoil pressure side is given for angles of attack lower than 5° which is translated into a gain of lift and an increase of drag. The flow also separates in the 25% thick airfoil but for further negative angles of attack which are redundant for the analysed cases of power production.

It is demonstrated that ΔAEP slightly varies with blade thickness for smooth blades. Additionally, a DEL reduction due to blade thickness is also defined in clean conditions. As a result, the increase of blade thickness may be seen as a proper option for blade weight reduction. However, ΔAEP varies until 5% with respect to the thinnest blade under rough surface conditions. According to AEP loss, the 30% thick airfoil changes its tendency with blade thickness from linear to exponential until 22% of AEP loss. Finally, the optimization of K_λ and β_{min} parameters demonstrates a reduction of roughness impact on AEP loss.

The probable operational angle of attack range is demonstrated to be dependent on the surface roughness state. A damaged blade surface decreases the angle of attack range from the one defined in a smooth surface. Consequently, the probability of flow separation is increased due to the augmentation of the blade thickness.

For future blade designs, a balance should be found between the gain in stiffness and the risk of having flow separation consequences as well as AEP losses due to thick airfoil presence.

References

[1] Nielsen M. S. and Nikolov I. and Kruse E. K. and Garnaes J. and Madsen C. B. 2020 High-resolution structure-from-motion for quantitative measurement of leading-Edge roughness *Energies* **13** 3916.

[2] Kruse E. K. and Sørensen N. and Bak C. and Nielsen M. S. 2020 CFD simulations and evaluation of applicability of a wall roughness model applied on a NACA 633-418 airfoil *Wind Energy* **23** 2056–2067.

[3] Smil V. 2019 Wind Turbines Just Keep Getting Bigger, But There's a Limit. *IEEE Spectrum.*

[4] Jensen F. and Branner K. 2013 *Advances in Wind Turbine Blade Design and Materials* Woodhead Publishing Series in Energy: Number 47 (Cambridge UK) chapter 1 p 3

[5] Timmer W.A. and Bak C. 2013 *Advances in Wind Turbine Blade Design and Materials* Woodhead Publishing Series in Energy: Number 47 (Cambridge UK) chapter 4 p 118-146

[6] Abbott I.H. and von Doenhoff A.E. 2012 *Theory of Wing Sections: Including a Summary of Airfoil Data* Dover Publications (New York).

[7] Van Rooij R. P. J. O. M. and Timmer W. A. 2003 Roughness sensitivity considerations for thick rotor blade airfoils *J. Sol. Energy Eng. Trans. ASME* **125**(4) 468-478.

[8] Gutiérrez R and Llorente E. and Echeverría F. and Ragni D 2020 Wind tunnel tests for vortex generators mitigating leading-edge roughness on a 30% thick airfoil *J. Phys. Conf. Ser.* **1618** 052058.

[9] Bak C. and Forsting A. M. and Sorensen N. N. The influence of leading edge roughness, rotor control and wind climate on the loss in energy production *J. Phys. Conf. Ser.* **1618** 052050.
[10] Jonkman J. and Butterfield S. and Musial W. and Scott G. 2009 *Definition of a 5-MW Reference Wind Turbine for Offshore System Development* Technical Report NREL/TP-500-38060.
[11] Popko W. et al. 2018 *IWES Wind Turbine IWT-7.5-164 Rev 4* Fraunhofer Institute for Wind Energy Systems IWES. Bremerhaven.
[12] Bak C. et al. 2013 *Description of the DTU 10 MW Reference Wind Turbine.* DTU Wind Energy Report-I-0092
[13] Piegl L and Tiller W. 1995 *The NURBS Book. 2nd ed.* Springer Berlin, Heidelberg.
[14] Boor, C. 2001 *A Practical Guide to Splines. Applied Mathematical Sciences* Springer (New York).
[15] IEC 61400-1 Wind Turbines - Part 1: Design Requirements Edition 4 *IEC International Standards. International Electrotechnical Commission.*
[16] Gunjit S. B. 2005 User's Guide to PreComp *Tech. rep. Golden, CO: National Renewable Energy Laboratory.*
[17] Garrad Hassan & Partners Ltd Bladed Theory Manual Version 4.3 St Vincent's Works, Silverthorne Lane, Bristol BS2 0QD England.
[18] Farin G 2001 *Curves and Surfaces for CAGD. A Practical Guide 5th Edition.* Morgan Kaufmann Publishers San Francisco CA United States.
[19] Anderson J.D. 2010 *Fundamentals of Aerodynamics* McGraw-Hill Education (New York).
[20] Pires O and Munduate X and Boorsma K and Ceyhan O and Madsen H A. and Timmer W.A. 2018 Experimental investigation of Surface Roughness effects and Transition on Wind Turbine performance *Journal of Physics: Conf. Series* **1037** 052018.
[21] Mayes C. and Schlichting H. and Krause E. and Oertel H.J. and Gersten K. 2003 *Boundary-Layer Theory* Springer (Berlin, Heidelberg).
[22] Nikuradse, J. 1950 *Laws of flow in rough pipes* Tecnichal Memorandum 1292 National 26 Advisory Committe for Aeronautics Washington (USA).
[23] Bak C. and Gaunaa M. and Olsen S.A. and Kruse E. K. 2016 What is the critical height of leading edge roughness for aerodynamics? *J. Phys.: Conf. Ser.* **753** 022023.
[24] Mendez B and Muñoz A and Pires O and Munduate X 2017 *Characterization of the carborundum used in rough airfoil surface tests and modelling with CFD* 35th Wind Energy Symp. (Grapevine, Texas) 0916.
[25] Gaudern N. 2014 A practical study of the aerodynamic impact of wind turbine blade leading edge erosion *Journal of Physics: Conf. Series* **524** 012031.
[26] Wilcox B. and White E. 2016 Computational analysis of insect impingement patterns on wind turbine blades *Wind Energy* **19** pp 483–49.
[27] Krishnan K. and Robison R. and Tetteh E. and Loth E. and Farrell T. and Crouch J. and Berry D. 2016 *Characterization of insect residue on an aerodynamic leading edge* 8th AIAA Atmospheric and Space Environments Conference (Washington, D.C.) p 3445.
[28] Papi F and Balduzzi F and Ferrara G and Bianchini A 2021 Uncertainty quantification on the effects of rain-induced erosion on annual energy production and performance of a Multi-MW wind turbine. *Renewable Energy* **165** pp 701-715.
[29] Maniaci C. D. and Westergaard C. and Hsieh A. and and Paquette A. J. 2020 *J. Phys.: Conf. Ser.* **1618** 052082

Estimation of rotor and main bearing loads using artificial neural networks

A. Loriemi, G. Jacobs, D. Bosse

CWD – Chair for Wind Power Drives
RWTH Aachen University
Campus-Boulevard 61
52074 Aachen

amin.loriemi@cwd.rwth-aachen.de

Abstract. Wind energy is one of the most important technologies for a climate-neutral energy supply. However, the premature failure of wind turbines due to unknown loads leads to a reduction in competitiveness compared to other energy sources. Here, load monitoring systems can make a significant contribution to the prevention of such failures. Most load monitoring systems for wind turbines focus on strain signals of structural components as the tower, main shaft or the rotor blade root. Based on these signals, axial forces, torsion or bending moments, which are acting on these components, are calculated. But this provides only partial information about the complete load situation, as transverse forces are not considered. Other methods use simulation or measurement data to train artificial neural networks (ANN) to estimate damage-equivalent loads acting on a wind turbine. However, this approach is accompanied by a loss of information, because the individual load components are condensed to an equivalent. In this work a method is presented that enables a measurement of rotor and main bearing loads considering all their individual load components. For this purpose, an ANN is trained with elastic multibody simulation (eMBS) data. Based on displacement signals, acting rotor and main bearing loads are estimated. The results show that even with consideration of nonlinearities, including nonlinear stiffness curves and bearing clearances, an appropriate accuracy can be achieved using the method presented.

1. Introduction and Objective

Over the last decades the size of wind turbines continuously increased [1]. The rising rotor diameter results in higher non-torque loads, which are acting on the rotor suspension system. Monitoring these loads provides many possibilities, especially to reduce the operating costs. By considering non-torque loads within the control strategy for example, the remaining useful lifetime of wind turbines can be extended significantly [2]. Based on the load history, maintenance intervals can be adapted and, if necessary, an early replacement of components can be initiated. Unfortunately, current load monitoring systems do not contain a detailed measurement of non-torque loads. Most load monitoring systems focus on strain signals of structural components as the tower, main shaft or the rotor blade root. Based on these signals, axial forces, torsion or bending moments, which are acting on these components, are calculated [3, 4]. But this provides only partial information about the complete load situation, as transverse forces are typically not measured using this technology and in the case of blade roots these methods are also subject to uncertainties due to thermal effects and nonlinear material properties [5, 6]. Additionally, classic strain gauges suffer from signal drift [7, 8] and have a limited lifetime of typically 1 to 3 years

[8], which makes this technology inappropriate for a long-term application. Other methods use simulation or measurement data to train artificial neural networks (ANN) to estimate damage-equivalent loads acting on a wind turbine [9, 10]. However, the conversion to damage-equivalent loads means that the individual load components, such as bending moments and transverse forces, can no longer be differentiated, which complicates the application within a turbine control system or a later understanding of harmful load situations.

In this work a method for a measurement of rotor and main bearing loads is presented, which enables an adequate consideration of the individual load components. An ANN is trained with elastic multibody simulation (eMBS) data for this purpose. In contrast to other methodologies, load estimation is based only on displacement signals of the drivetrain. These kinds of signals are generally simple to access, as the corresponding sensors can be mounted quickly and easily, even when the wind turbine is in operation. In addition, in the case of non-contact sensors, a wear- and maintenance-free long-term operation of the sensor system is possible. It is analyzed what accuracy can be achieved with this methodology, even if measurement noise is considered. Further the necessary capability of ANNs for rotor and main bearing load estimation are evaluated, including the representation of nonlinearities and dynamic system behavior.

2. Methods

The target functionality of the method is to estimate rotor and main bearing loads only using displacement signals. Unfortunately, displacement signals can be subject to nonlinearities, which complicates a direct conversion of the signals into acting loads. Physical models which describe the correlation between loads and displacements often cannot be solved directly. Instead, these types of models require an iterative solving process, which computational cost can prevent a real-time application. Additionally, physical models show a less robust behaviour to measurement errors and deviation of system parameters as bearing clearances. To overcome these problems a data-driven modelling approach is chosen (Figure 1).

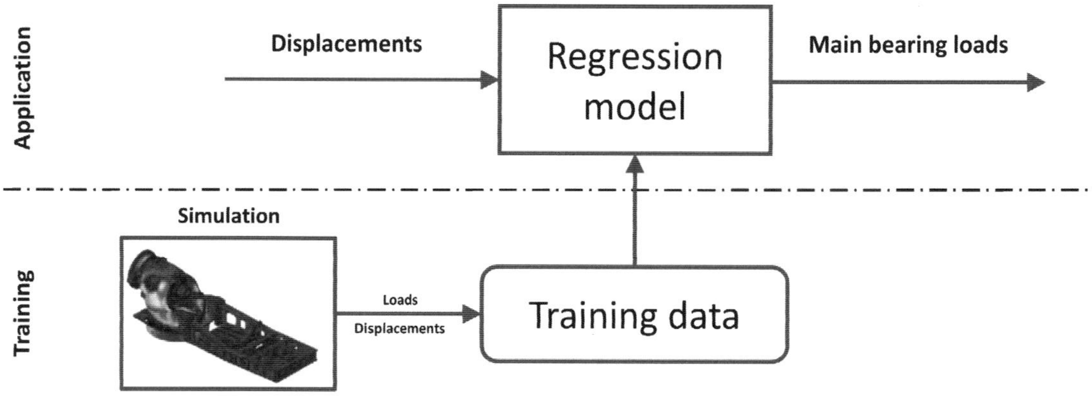

Figure 1. Displacement based estimation of rotor and main bearing loads

Based on training data, regression models are developed, which enable a direct calculation of rotor and main bearing loads and with that a real-time application. Furthermore, this type of modelling offers the possibility to increase the robustness against measurement errors by considering a distortion of the input signals within the training process. However, the performance of a regression model depends heavily on the quality of the training data. The training data has to be representative for the system response as only included correlations can be approximated reliably. A cost-effective way to generate such data is simulation. In order to do this, it is necessary to represent the system with a sufficient model depth considering the physical effects, which have a significant influence on the system response.

As a prove of principle study, a nonlinear and dynamic simulation model of a wind turbine is developed, which provides a very detailed model depth. Operational rotor and main bearing loads are estimated by using resulting displacement signals of the wind turbine model. In this way, the accuracy is determined that is basically achievable with the presented approach. Furthermore, the requirements regarding the representation of nonlinearities and dynamics are evaluated.

2.1. Simulation model

The analysis of this study refers to a 2.3 MW low wind speed turbine (rotor diameter: 141m; nominal wind speed: 9 m/s) that has been developed in the context of the project "MaxCap" [11]. The rotor suspension system is designed as a four-point suspensions, as it is shown in Figure 2.

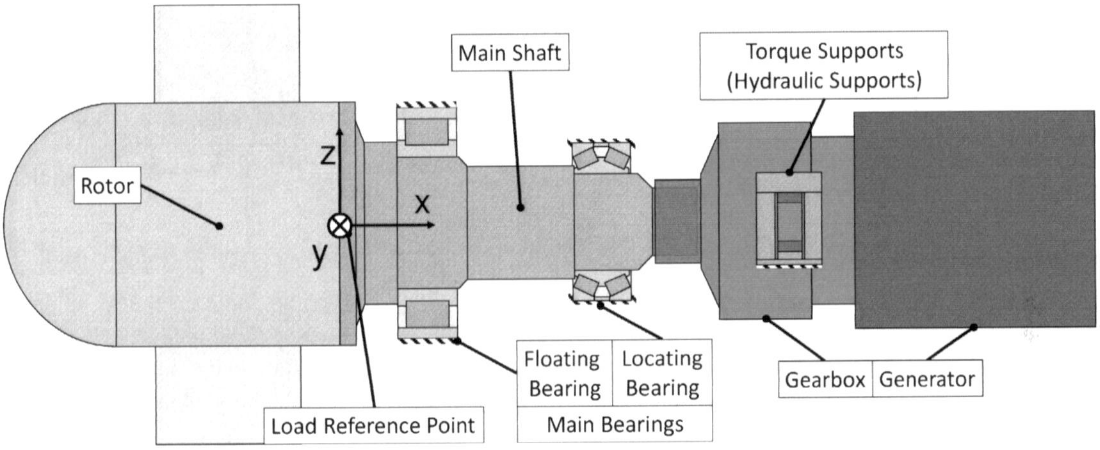

Figure 2. Principle sketch of considered four-point suspension

A cylindrical roller bearing (CRB) as a floating bearing and a double row tapered roller bearing (TRB) as a locating bearing are used as main bearings. The CRB has a radial clearance, whereas an axial clearance is specified for the TRB. Both clearances are dependent on temperature, manufacturing and mounting and therefore are subject to scatter. Bearing clearances have a significant influence on the resulting displacements, since they lead to a discontinuity of the bearing stiffness curve. A simulation model that is intended to represent the relationship between displacements and loads must take this relationship into account.

The gearbox is connected to the machine carrier by a hydraulic suspension. This type of torque supports is characterized by a high torsional stiffness, whereas a translational movement of the gearbox in relation to the machine carrier is only met with a low resistance [12]. Thereby, the rotor input torque is supported only by this component, leaving non-torque loads being supported mostly by the main bearings. Since its functionality is realized by fluid-filled elastomer chambers, this component has a strongly nonlinear stiffness behaviour, which must be considered by an appropriate modelling.

As required by the current guidelines for load calculation, the rotor and tower dynamics must be represented in the model. A simulation model must therefore include an aero elastic load calculation and take into account the flexibility of structural components. Otherwise, the simulation results in wrong structural dynamics and loads.

To meet these requirements a detailed eMBS-model of wind turbine has been developed using the multi body simulation software SIMPACK (Figure 3).

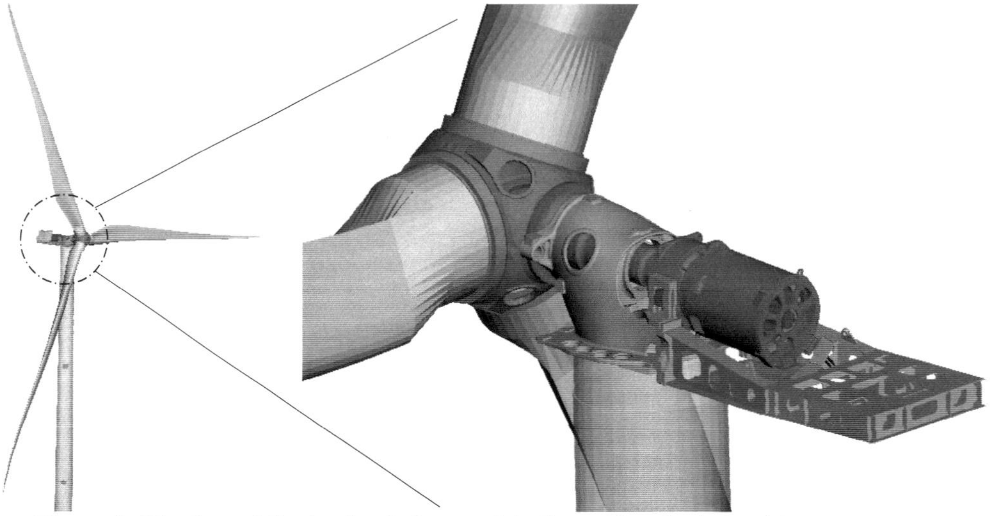

Figure 3. Elastic multibody simulation model of considered wind turbine

Inside this model, relevant structural components are represented as flexible bodies so that structural-dynamic effects are taken into account, when calculating the system response. This includes following components: rotor blades, hub, tower, main shaft, main bearing housing, machine carrier, gearbox housing, and planet carriers. The elastic properties of the bearings in the drivetrain are taken into account by lumped-parameter roller bearing models in their software implementation. In this implementation, the bearing reaction forces are calculated using algebraic equations [13, 14] on basis of bearing ring kinematics and a lamina roller model [14]. Using lamina roller models, the roller is discretised into multiple slices along the roller axis. The contact stiffness is then calculated for every slice of the roller, which enables the consideration of a roller profile and bearing clearance. Especially the consideration of the bearing clearance is necessary to calculate the correct bearing ring displacements, as these variables are directly influenced by this parameter. Within the wind turbine model, this approach was used to model the main bearings and planet carrier bearings. Thereby, parameter values have been determined using the manufacturers' specifications. The stiffness and damping characteristics of the torque supports were determined by experiments on a component test bench and implemented as nonlinear spring-damper elements. Loads acting on the rotor blades are calculated in respect to local wind speed and wind direction given by turbulent wind fields generated with TurbSim [15]. This is done using the blade element theory in its implementation in the AeroDyn code [16]. Table 1 shows the specification of the simulations performed.

Table 1. Specification of training and testing data

	Wind speed (incrementation)	Radial clearance deviation at floating bearing (incrementation)	Axial clearance deviation at locating bearing (incrementation)
Training Data	3 m/s to 24 m/s (1 m/s)	0 μm (-)	0 μm (-)
Testing Data	6 m/s to 12 m/s (3 m/s)	0 μm (-)	0 μm (-)
Sensitivity Analysis	6 m/s to 12 m/s (3 m/s)	-100 μm to 100 μm (50 μm)	-150 μm to 150 μm (25 μm)

For training data, the wind speed is varied from 3 m/s (cut-in speed) to 24 m/s (cut-out speed), while clearances of the main bearings are kept constant in accordance with their nominal values. Thereby a wide load range of the wind turbine is covered by the training data. Constant clearances and three wind speeds, covering different operation modes (partial, nominal and full load), are considered for testing. In contrast to that, a clearance deviation based on the estimated scatter of these parameters is considered within the data for sensitivity analysis. Here, one clearance is varied while the other is kept constant at its nominal value. In this study, clearance deviation is considered only for the main bearings as most of the displacement signals used for load estimation are given locally at these components. Thereby the estimation is assumed to be most sensitive to these parameters.

Output variables of the simulations are the six rotor loads (Fx, Fy, Fz, Mx, My and Mz) acting at the main shaft's rotor flange. Their location and orientation are given by the load reference point and the corresponding coordinate system shown in Figure 2. Additionally, the main bearing loads are calculated including two radial forces of the floating bearing ($Fy_{,FB}$ and $Fz_{,FB}$) and one axial and two radial loads of the locating bearing ($Fx_{,LB}$, $Fy_{,LB}$ and $Fz_{,LB}$). Displacements are calculated for the three translational displacements and two tilting displacements (relative motion of inner ring to outer ring) for each main bearing. Additionally to these 10 variables, the three translational displacements at each gearbox torque arm at the torque supports are used, resulting in a total of 16 displacement signals. In a post-processing procedure, according to estimated sensor uncertainties, the calculated displacement signals are superimposed with an artificial normal distributed measurement noise. Here, the measurement noise is modelled as a normally distributed error, which is added to the actual simulation result. The distortion of the signals is necessary, because otherwise the chosen method would achieve an unrealistic estimation accuracy that could not be achieved under real measurement conditions. In order to evaluate the method in terms of its applicability, the measurement error must therefore also be simulated.

2.2. Regression models
On the basis of the noisy displacement signals and the corresponding rotor and main bearing loads, different regression models are trained. As the simplest regression model, a linear regression (LinReg) of input and output values is performed. This model type thereby approximates the considered system with a static and linear behaviour. Even if it is known in advance, that the system has nonlinear and dynamic characteristics, this rather unsuitable regression model provides a benchmark for other regression models.

To model even strongly nonlinear correlations between displacements and loads, ANNs are chosen as further regression models. In a first approach, feedforward neural networks (FNN) are trained to estimate rotor and main bearing loads. FNNs enable the consideration of nonlinearities but still only approximate a static behaviour of the system. Additionally to FNNs and to overcome this abstraction loss, long short-term memory (LSTM) layers are used to expand an ANN topology. LSTM layers give the ANN the functionality to store information from previous time steps of sequence data and enable the consideration of time-dependent effects. Therefore a nonlinear and dynamic representation of the system is achieved with LSTM networks.

In this study, linear regression models, FNNs and LSTM networks are compared with each other. For all ANNs, the training data is split into two data sets. Thereby 180,000 data samples are used for actual training and 84,000 data samples for validation. For both, FNNs and LSTM networks, training is ended when no further improvement in accuracy is achieved for the validation data set. In parallel, linear regression is performed on the same training data used to train the ANNs.

Different parameters are calculated to evaluate the individual regression models. A first parameter is the Root Mean Square Error (RMSE)

$$RMSE = \sqrt{\sum_{i=1}^{N} (y_i - \hat{y}_i)^2 / N} \tag{1}$$

where y_i describes the calculated load, $\hat{y}_i$ the estimated load and N the number of samples in the testing data. This parameter evaluates the absolute estimation error of the individual load component. Based on this parameter the relative RMSE is calculated.

$$Relative\ RMSE = \frac{RMSE}{Load\ Range} \tag{2}$$

This parameter sets the RMSE in context to the load range of the testing data. Thereby it can be evaluated, if the estimation error is sufficient for the planned application. Additionally, to quantify the quality of the regression itself, the coefficient of determination is calculated.

$$R^2 = 1 - \frac{\sum_{i=1}^{N}(y_i - \hat{y}_i)^2}{\sum_{i=1}^{N}(y_i - \bar{y})^2} \tag{3}$$

where $\bar{y}$ is the mean of the load within the testing data.

3. Results

Training of the linear regression model, FNN and LSTM network has been performed with simulation data without any variation of main bearing clearances according to the specification of Table 1. The load range of the calculated rotor and main bearing loads covered by the training data is used to calculate the relative RMSE. The RMSE is determined based on the difference between calculated and estimated loads. The trained models are used to estimate the loads of the testing data (Table 1) where bearing clearances are kept the same as for training and no deviation occurs. In Table 2 the evaluation parameters are shown for the estimation of calculated rotor loads using the displacement signals of the simulation model. Similarly, Table 3 shows the results for the estimation of the calculated main bearing loads.

Table 2. Results for rotor loads and constant bearing clearances

		Fx	**Fy**	**Fz**	**Mx**	**My**	**Mz**
	Load Range	550 kN	220 kN	220 kN	3.2 MNm	10.8 MNm	9.6 MNm
Absolute RMSE	LinReg	35 kN	9 kN	12 kN	31 kNm	51 kNm	49 kNm
	FNN	24 kN	7 kN	11 kN	13 kNm	42 kNm	32 kNm
	LSTM	22 kN	8 kN	11 kN	9 kNm	45 kNm	32 kNm
Relative RMSE	LinReg	6,3%	4,2%	5,5%	1,0%	0,5%	0,5%
	FNN	4,3%	3,4%	5,0%	0,4%	0,4%	0,3%
	LSTM	4,1%	3,6%	5,2%	0,3%	0,4%	0,3%
R^2	LinReg	84,01%	48,39%	52,06%	99,69%	99,84%	99,76%
	FNN	93,99%	58,83%	61,49%	99,94%	99,85%	99,89%
	LSTM	93,97%	59,92%	60,65%	99,96%	99,87%	99,88%

Table 3. Results for main bearing loads and constant bearing clearances

		$Fy_{,FB}$	**$Fz_{,FB}$**	**$Fx_{,LB}$**	**$Fy_{,LB}$**	**$Fz_{,LB}$**
	Load Range	5.6 MN	6.2 MN	650 kN	5.5 MN	5.9 MN
Absolute RMSE	LinReg	33 kN	34 kN	34 kN	31 kN	31 kN
	FNN	19 kN	27 kN	23 kN	18 kN	23 kN
	LSTM	19 kN	25 kN	21 kN	20 kN	22 kN
Relative RMSE	LinReg	0,6%	0,5%	5,2%	0,6%	0,5%
	FNN	0,3%	0,4%	3,5%	0,3%	0,4%
	LSTM	0,3%	0,4%	3,2%	0,4%	0,4%
R^2	LinReg	99,62%	99,79%	86,41%	99,70%	99,81%
	FNN	99,86%	99,86%	95,40%	99,89%	99,89%
	LSTM	99,83%	99,87%	95,40%	99,87%	99,89%

FB: Floating Bearing
LB: Locating Bearing

It can be stated, that without any deviation in main bearing clearances, ANNs show a more accurate load estimation compared to a linear regression model. Using ANNs, especially radial forces of the main bearings ($Fy_{,FB}$, $Fz_{,FB}$, $Fy_{,LB}$, $Fz_{,LB}$), torque (Mx) and bending moments (My, Mz) at the rotor flange can be estimated with a relative RMSE of less than 1%, referring to the load range to be covered. However, rotor thrust (Fx), transverse forces at the rotor flange (Fy, Fz) and axial forces of the main bearing arrangement ($Fx_{,LB}$) are associated with a significantly higher relative RMSEs, but mostly less than 5%. Comparing FNNs and LSTM networks it is shown that no significant improvement can be achieved with the exception of the rotor input torque Mx. Here the RMSE can be reduced by 30 % by using LSTM networks instead of FNNs. A similar pattern arises for the coefficients of determination. Values of more than 99 % are achieved for torque and bending moments at the rotor flange and radial forces of the main bearings. Rotor thrust, axial main bearing forces and especially transverse forces at the rotor show a worse coefficient of determination.

To evaluate the sensitivity of the regression models to clearance deviations, the corresponding data set (Table 1) is analysed. Within this data set, the radial clearance of the floating bearing and the axial clearance of the locating bearing is varied. The resulting RMSE quantifies the error of estimating the calculated loads using regression models which do not include this variation within the training data. The estimation errors for rotor loads, depending on the clearance deviation are shown in Figure 4.

 doi:10.1088/1742-6596/2151/1/012002

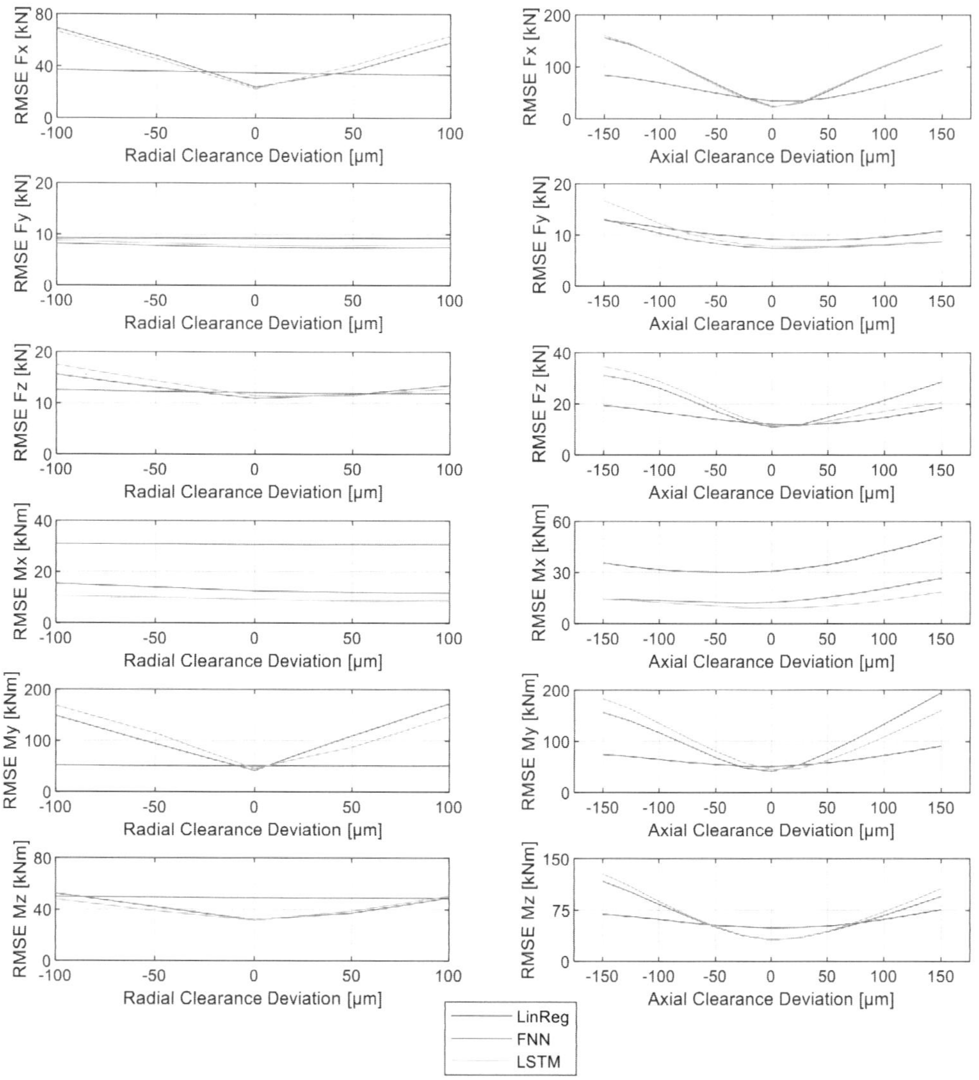

Figure 4. RMSE for rotor load estimation depending on radial clearance deviation of floating bearing (main bearing arrangement) and axial clearance deviation of locating bearing (main bearing arrangement)

The clearance deviations do affect the individual load components differently. For rotor thrust Fx, the accuracy of the ANNs can only be kept in a small range of clearance deviation. Outside of this range the estimation error increases drastically, whereby the linear regression model shows a more robust behaviour. A similar statement can be made for estimating the transverse force Fz and bending moment My. In contrast to that, the estimation of the transverse force Fy, torque Mx and bending moment Mz using ANNs show a better accuracy than using a linear regression model over a wide range of clearance deviation. Analogously, the estimation errors for main bearing loads are shown in Figure 5.

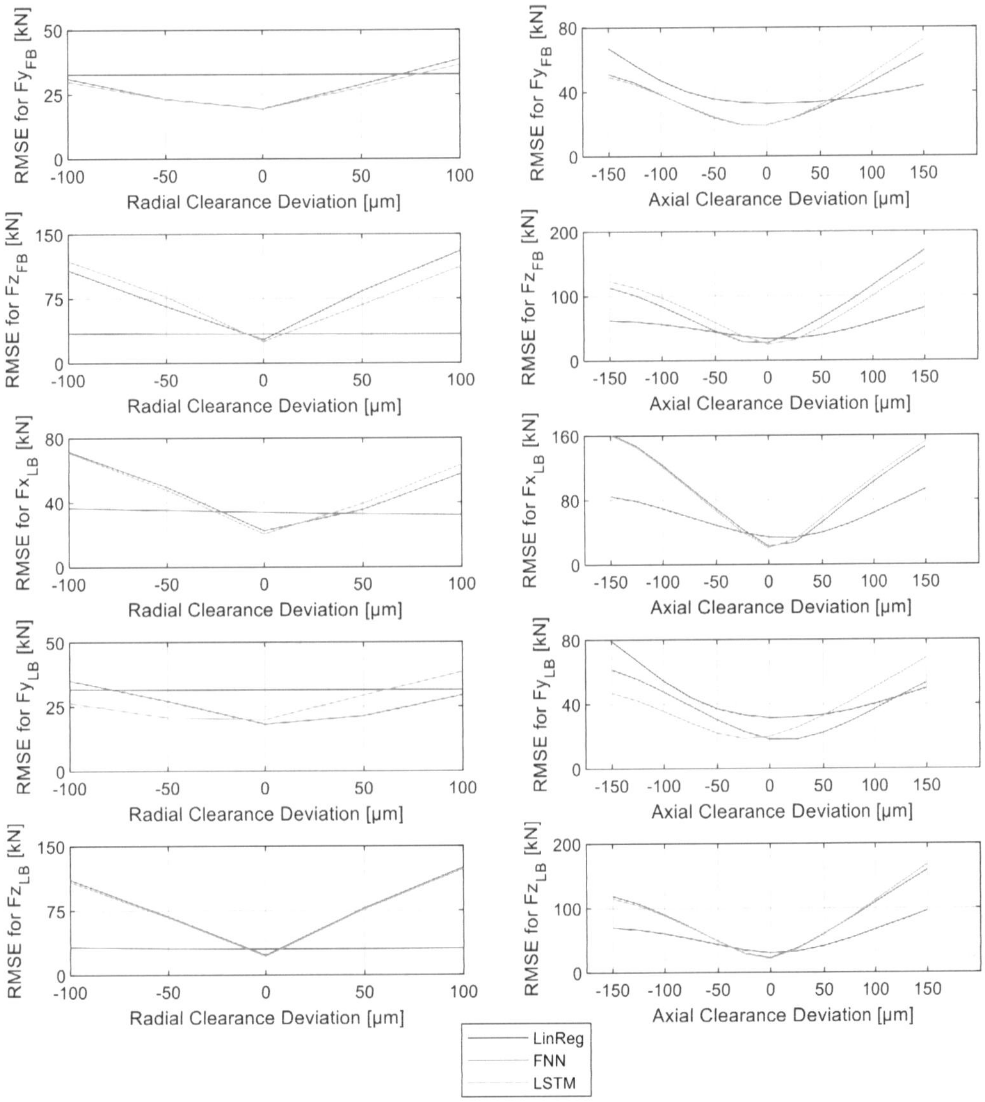

Figure 5. RMSE for main bearing load estimation depending on radial clearance deviation of floating bearing (main bearing arrangement) and axial clearance deviation of locating bearing (main bearing arrangement)

As with the estimation of the rotor loads, the estimation of main bearing loads can component-wise be separated in two groups. Regarding radial forces $Fz_{,FB}$ and $Fz_{,LB}$ as well as the axial force $Fx_{,LB}$, the accuracy of ANNs has its validity only for minor clearance deviations. Outside a rather small deviation range, ANNs are outperformed by simple linear regression models. In contrast to that, estimating the radial forces in y-direction ($Fy_{,FB}$ and $Fy_{,LB}$) using ANNs shows a superior accuracy, even though there is a wider deviation of main bearing clearances.

4. Discussion

The results show that the approach of estimating rotor and main bearing loads using regression models is reasonable and promises a good accuracy. Training and testing data have been generated using a detailed state-of-the-art eMBS model, what promises a good comparability with the real application. Even though measurement errors have been considered, relative RMSE of the individual loads of no more than 5 % are achieved. The fact that the accuracy of ANNs is better than the accuracy of linear regression models proves that the nonlinearities of the system should be taken into account by the simulation model used for training data generation. Otherwise these nonlinearities cannot be approximated by a regression model, what lowers the accuracy.

It is outstanding, that for most of the considered loads, a dynamic representation of the system does not increase the estimation accuracy significantly. In particular, the use of an LSTM network decreased the RMSE of the rotor input torque estimation. This effect is comprehensible, since the estimation of this load is mainly driven by the displacements of the torque supports. This component has relevant damping properties, which are accompanied by a corresponding hysteresis of the torque arm displacements. Therefore, it is understandable that a time series-based estimation provides a better accuracy for this load. However, for all other loads, a static representation of the system appears to be sufficient for load estimation.

The sensitivity analysis has shown partial disadvantages of ANNs due to varying bearing clearances. The estimation of certain load components using ANNs show a high sensitivity to this system parameters and are even outperformed by the use of linear regression models. This can be explained by the fact that the ANNs were trained and optimized to have the best accuracy for the system represented by the training data. In contrast, the systems represented in the data for sensitivity analysis have bearing clearances whose deviations are not included in the training data. The results show that the significant accuracy of the ANNs for the original system is not transferred to these deviating systems. Considering the less accurate but more robust properties of a linear regression, an opposing relationship between accuracy and robustness is indicated. Accordingly, a trade-off between these two attributes may has to be made. One solution strategy for this problem could be the inclusion of a clearance scatter within the training data. The ratio between accuracy and robustness could then be controlled by the width of the scatter considered in the training data. Alternatively, the regression model could be changed. For instance, nonlinearities could also be represented by polynomial regression models, where the ratio of accuracy and robustness could be adjusted by the degree of the polynomial.

5. Conclusions and Outlook

With the approach presented, it was possible to estimate rotor and main bearing loads considering the individual components of the load vector. This is done by using ANNs which are trained with displacement signals of the drivetrain. This kind of measurement signals are easy and cheap to record, which makes this method practical for commercial use. Since measurement uncertainties and nonlinearities were taken into account, it can be assumed that similar results can also be achieved for the real system of a wind turbine. Consequently, the next step must be a transfer of the methodology applied here to the real system of a wind turbine, in order to prove its usability in practice.

Nevertheless, a sensitivity of the estimation accuracy to varying system properties could be shown. The scatter of system parameter like bearing clearances can massively decrease estimation accuracy if it is not considered by the estimation model. Future research must therefore also develop solution strategies on how to achieve the necessary balance between accuracy and robustness considering scattered system parameters. It is becoming apparent that load-component-specific solutions promise better results than a general solution approach. For example, it might be more appropriate to use a linear regression model to estimate rotor thrust, while at the same time an ANN might be better suited to estimate rotor bending moments. However, in order to find a sensible trade-off, it will be necessary in the future to specify what kind of accuracy or robustness is required, which can vary strongly depending on the load monitoring application.

6. Acknowledgements
The authors would like to thank the Ministry of Economic Affairs, Innovation, Digitalization and Energy of the State of North Rhine-Westphalia, Germany, for the financial support granted. They also thank their project partners, and in particular windwise, for providing the equipment, insight and expertise that contributed significantly to this successful joint research.

EUROPÄISCHE UNION
Investition in unsere Zukunft
Europäischer Fonds
für regionale Entwicklung

EFRE.NRW
Investitionen in Wachstum
und Beschäftigung

References

[1] Enevoldsen P and Xydis G 2019 *Energy Sustain. Dev.* **50** 18-26
[2] Njiri J G, Beganovic N, Do M H and Söffker D 2019 *Renew. Energ.* **131** 818-28
[3] IEC 61400-13:2015, Wind turbines - Part 13: Measurement of mechanical loads
[4] Lekou D J and Mouzakis F 2009 *J. Sol. Energy Eng.* **131** 011005
[5] Papadopoulos K, Morfiadakis E, Philippidis T P and Lekou D J 2000 *Wind Energy* **3** 35-65
[6] Bezziccheri M, Castellini P, Evangelisti P, Santolini C and Paone N 2017 Wind Energy **20** 1997-2010
[7] Berg J C, Resor B R, Paquette J A and White J R 2014 *SMART Wind Turbine Rotor: Design and Field Test,* Sandia National Laboratories, Report No.: SAND2014-0681
[8] Verbruggen T W, Braam H and Rademakers L W M M 2012 *Fibre Optic Blade Monitoring - Development of a cost effective system for O&M optimization*, Energy research Center of the Netherlands, Report No.: ECN-E—12-018
[9] Schröder L, Dimitrov N K, Verelst D R and Sørensen J A 2018 *J. Phys.: Conf. Ser.* **1037** 062027
[10] Cosack N 2010 *Fatigue Load Monitoring with Standard Wind Turbine Signals* (Aachen: Shaker) pp 95-104
[11] Abel D, Brecher C, De Doncker R W, Hameyer K, Jacobs G, Monti A, Schröder W and Hirt A 2019 *Conference for Wind Power Drives 2019: Conference Proceedings*, (Norderstedt: Books on Demand) pp 259-75
[12] Helsen J, Peeters P, Vanslambrouck K, Vanhollebeke F and Desmet W 2014 *Renew. Energ.* **69** 336-46
[13] Harris TA, Kotzalas M N 2006 *Advanced Concepts of Bearing Technology*, (Boca Raton: Taylor & Francis Group) pp 13-26
[14] Schlecht B 2010 *Maschinenelemente 2*, (Munich: Pearson Studium) pp 221-223
[15] NREL Homepage, https://www.nrel.gov/wind/nwtc/turbsim.html, last accessed 2020/09/03
[16] NREL Homepage, https://www.nrel.gov/wind/nwtc/aerodyn.html, last accessed 2020/09/03

Generating long-term sub-hourly wind speed time series by coupling mesoscale models with full-scale spectra

David Schillebeeckx[1] and Grégoire Leroy[1]
[1] 3E nv, Kalkkaai 6, 1000 Brussels, Belgium
E-mail: david.schillebeeckx@3e.eu

Abstract. Time series-based wind studies are gaining more and more interest and importance within the wind resource community. Benefits of such time series-based analysis being, among others, more accurate production and loss estimations, better representation of ramp-up and extreme events, and improvements in the dimensioning of energy storage systems. While mesoscale Numerical Weather Prediction (NWP) models can simulate long-term winds that capture non-stationary weather patterns, it is known that they are not able to properly resolve sub-scale processes leading to a smoothing effect. Recent studies presented spectral models that describe the full atmospheric spectrum of boundary layer winds which connect the microscale turbulent movements to the large, mesoscale fluctuations. In this work, mesoscale simulations from the Weather Research and Forecasting (WRF) model are coupled with stochastic turbulence simulations using state-of-the-art full-scale boundary-layer spectra to efficiently bridge the spectral gap between the mesoscale and the turbulence fluctuations, without requiring any local measurements nor expensive CFD simulations. The study provides a practical step-by-step approach to generate long wind speed time series at high sampling rate at any desired location and height.

1. Introduction

As an atmospheric quantity, wind speed is naturally volatile and spatiotemporal fluctuations occur at a wide range of scales from planetary and annual variations down to the turbulent scale. For wind energy yield assessments, these wind variations have historically been represented through statistics and histograms [1]. However, with the increase of computational power, wind energy production estimates are more and more being calculated directly in the time domain. The use of time series typically lowers the uncertainty as it avoids imposing Weibull distributions that introduce fitting errors [2]. Additionally, time series include the atmospheric seasonal and diurnal variations into the production estimations, hereby also introducing the varying atmospheric quantities such as stability, air density, turbulence and wind shear into the calculations allowing to correctly model their influence on the turbines performance and turbine interaction effects. Time series-based calculations also enable to estimate complex time-dependent losses and the interaction and overlap between them more accurately. These include for example losses due to bat and bird protection stops, noise reduction operations modes, grid export limitations, shadow flicker curtailments, etc. Consequently, the combination of these benefits of time series-based energy yield assessments allows to lower the uncertainty on the long-term production estimation. Additionally, for turbine load assessments, high-frequency wind data down to the turbulent scale are imperative to capture the dynamic effects of the turbine and estimate the components fatigue life [3]. Furthermore, the increasing share of intermittent wind power plants in the energy production mix also calls for wind generation time series with a high temporal resolution. For high penetration of wind in the power system, sub-hourly production data for wind integration studies are required to correctly capture the load variability, start-ups and ramp effects [4]. High resolution time series are also crucial for complex wind and storage systems, as the time resolution of the data can significantly impact the dimensioning and investment decision of storage systems [5]. Lastly, from a financial point of view, balancing of the consumption and production profiles is becoming more challenging with volatile electricity prices and complex hybrid plants combining different production and storage technologies.

Numerical Weather Prediction (NWP) models, such as the mesoscale Weather Research and Forecasting (WRF) model, and reanalysis products can provide long-term wind time series based on historical weather data that are proven to be a reliable source for wind energy assessment studies [6, 7, 8]. Although numerical simulations show great performances for the estimation of the mean wind regime over large areas, the models are constrained by their spatial and temporal resolution limits. The terrain is represented by a relatively coarse grid with a resolution of a few kilometres and therefore does not allow to capture the smaller sub-grid phenomena, such as topographic speed-up effects [9]. Furthermore, because of the filtering and dampening numerical schemes in mesoscale models, the effective temporal resolution is further reduced [10]. These limitations of the effective model resolution lead to a smoothing effect of the wind and therefore to a reduced wind variability [9, 11]. The spatial and temporal averaging of mesoscale simulations result in low-resolution, e.g. hourly, wind time series that do not capture the actual microscale fluctuations and turbulence on site. Downscaling these low-resolution wind simulations using high fidelity Computational Fluid Dynamics (CFD) models is possible but comes at a high computational cost [12]. While more cost-efficient methods exist to spatially downscale mesoscale simulations and correcting for the coarse topographic representation such as the WRF-WAsP method in [13], efficient alternatives for a temporal downscaling, going from hourly to a higher temporal resolution, are lacking. The CorRES tool [14] introduces sub-hourly variations to hourly WRF simulations with stochastic model that simulates high frequency fluctuations from a power spectrum density (PSD) function. However, similar as in the Spectral Correction Method (SCM) from [9, 11], the spectrum for higher frequencies is extended following $S(f) \propto f^{-5/3}$ which is only valid for lower frequencies and therefore limits its applicability to model fluctuations up to $\sim 10\,\mathrm{min}$. Recent studies present full-scale boundary layer spectra that cover the mesoscale atmospheric motions down to the fast fluctuations of the boundary layer turbulence [15, 16]. Therefore, these spectral models enable to simulate wind speed fluctuations for a higher frequency and generate realistic turbulence that can be added to the mesoscale time series.

This paper presents such a hybrid approach by coupling mesoscale wind speed with stochastic turbulence simulations based on full-scale boundary layer models. Similar principles were applied as in [17], where measurements with a 10 min resolution were used to capture the non-stationary wind speed characteristics instead of NWP simulations. The authors also mention the possibility of using mesoscale simulations as an alternative for measurements but address the limitation of effective time resolution of such models. This work addresses this issue by first applying a correction on the mesoscale fluctuations in a similar fashion as the SCM but directly on the spectral components rather than on the spectral moments to allow for a time series analysis (subsection 3.1). Then, physically realistic turbulence at any desired sampling are generated by building upon the advancements in the boundary layer spectral models from [15] and [16] (subsection 3.1). The combination of these modelled low- and high-frequency spectra covers the full boundary-layer spectrum and thus closes the spectral gap between the meso- and microscale (subsection 3.2). The methodology has been validated against observations in section 4. The mesoscale and measurement data used are briefly discussed in the following section and conclusions are drawn in section 5.

2. Data and background

To demonstrate the different steps of the methodology outlined in this paper, WRF simulations are used for the mesoscale input data. Similar model setup and parameterisations were used as in the New European Wind Atlas (NEWA) production run [8], but with an hourly output time step instead of half-hourly. Although WRF has been used for the purpose of showcasing the applied techniques, the methodology outlined here is generic and is suited for simulations from other NWP models or for reanalysis datasets as well.

As a reference, and for validation, measurements from a 30 m meteorological mast within the WRF domain were used. The measurements consist of 1-year of wind speed data from a cup anemometer with a recording frequency of 1 Hz. The site is land-based with a flat terrain, however, it can be considered as complex in term of roughness due to the presence of multiple industrial buildings an forest patches around the mast. Wind speed data has then been extracted from the closest grid point of the WRF simulation.

The reader should note that since the power spectra can appear relatively noisy at higher frequencies, logarithmic smoothing is often applied to remove this noise and highlight the shape of the spectrum.

3. Methodology

Mesoscale simulations show great performances in simulating wind speed time series that capture the non-stationarity of long-term variations. However, these simulations have two shortcomings as discussed in the introduction:

(i) the modelled spectrum underestimates the spectral energy level in the mesoscale range because of the dampening effect in the mesoscale simulations; and

(ii) the limited output time resolution of the simulations (typically hourly) do not capture the higher frequency fluctuations in the microscale range.

These two limitations of the mesoscale simulations will be tackled in three main steps. First, the dampening effect of the mesoscale simulations is alleviated by using the Larsén mesoscale spectral model [18] and correcting the modelled mesoscale time series in the frequency domain. As a next step, the low-resolution time series are temporally downscaled by adding physically realistic turbulence fluctuations using a recent full-scale spectral model [15, 16]. Finally, the resulting spectrum can be converted to the time domain with a Fourier transform to obtain high-frequency wind speed time series for any desired output resolution dt. This high frequency data can further be resampled to arrive at 10-minute mean and standard deviation wind speed values that mimics field observations in a realistic manner.

These steps are discussed in more detail in the following subsections.

3.1. Mesoscale spectral correction

As a first step, the smoothening effect by the WRF simulations that leads to an underestimation of the hourly wind variability will be corrected. The correction will be applied in the frequency domain and follows similar principles as the Spectral Correction Method (SCM) described in [9, 11].

The mesoscale wind behaviour can be represented with the Larsén mesoscale spectrum derived in [18]. This spectral model is applicable for the large-scale mesoscale wind fluctuations in the lower frequency range from approximately 10^{-6} Hz to 10^{-3} Hz:

$$S_{\text{Larsén}}(f) = a_1 f^{-5/3} + a_2 f^{-3}, \tag{1}$$

where the values of the two coefficients, $a_1 = 3 \times 10^{-4}\,\text{m}^2\,\text{s}^{-8/3}$ and $a_2 = 3 \times 10^{-11}\,\text{m}^2\,\text{s}^{-4}$, were obtained from fitting of the model to multiple measurements. To ensure that the spectral model matches the large-scale wind variability at the site, the Larsén mesoscale spectrum is calibrated to the specific site conditions to:

$$S_{\text{meso}}(f) = A \times S_{\text{Larsén}}(f). \tag{2}$$

The calibration coefficient A is obtained through regression from the observed mesoscale spectrum in the frequency range from 8×10^{-6} Hz to 3×10^{-5} Hz, where it is assumed that the WRF simulations are correctly capturing the spectral energy content.

The spectral correction is applied by scaling the WRF simulated spectral energy to the mesoscale spectral model of equation (2) for frequencies above a certain correction frequency f_c. Here, the best results were obtained by selecting $f_c = 2 \times 10^{-5}\,\text{Hz} \approx 1.7\,\text{d}^{-1}$, which is in line with the values reported in [9]. The correction is obtained by first applying a loglinear regression to the observed WRF spectrum:

$$S_{\text{linear}}(f \geq g_c) = b_1 f^{b_2}. \tag{3}$$

The ratio between this linear fit and the calibrated mesoscale spectral model of equation (2) then determines the scaling factor that will be applied to the WRF spectral components at each frequency while keeping the original WRF spectrum untouched for $f < f_c$:

$$S_{\text{WRF,corr}}(f) = \begin{cases} S_{\text{WRF}}(f) & f < f_c \\ S_{\text{WRF}}(f) \times \dfrac{S_{\text{meso}}(f)}{S_{\text{linear}}(f)} & f \geq f_c \end{cases} \tag{4}$$

As an example, and to visually represent the mesoscale correction, Figure 1 shows the WRF spectrum before (a) and after (b) the correction of equation (4) is performed. This correction effectively introduces higher wind variability to the original WRF simulations.

3.2. Microscale spectrum extension

The next step is to include the contribution of the sub-hourly turbulence effects to arrive at any desired output resolution dt $(= 1/f_{\text{output}})$. To simulate these microscale wind speed fluctuations in the high frequency range, the Mikkelsen-Tchen microscale boundary-layer spectral model as described in [19] has been considered. The two-sided representation of this model is:

$$S_{\text{Mikkelsen}}(f) = \frac{a u_*^2 n/n_l}{2f(1 + n/n_l)(1 + n/n_u)^{2/3}}, \tag{5}$$

where $n = fz/U$ is the normalized frequency at height z, u_* the friction velocity, $n_l = f_{\text{Cor}} z / 0.6 u_*$, $a = 0.953$ and $n_u = 0.185$.

To ensure a smooth transition between the meso- and the microscale, a full-scale boundary-layer spectrum model is required. For this, the approach developed in [15] and applied in [16] to superimpose the meso- and microscale spectral contributions is used:

$$S_{\text{meso+micro}}(f) = \text{equation (2)} + \text{equation (5)} = S_{\text{meso}}(f) + S_{\text{Mikkelsen}}(f). \tag{6}$$

The Fourier coefficients for the high frequency range are then generated from equation (6) using random phase angles following a similar methodology as described in previous studies

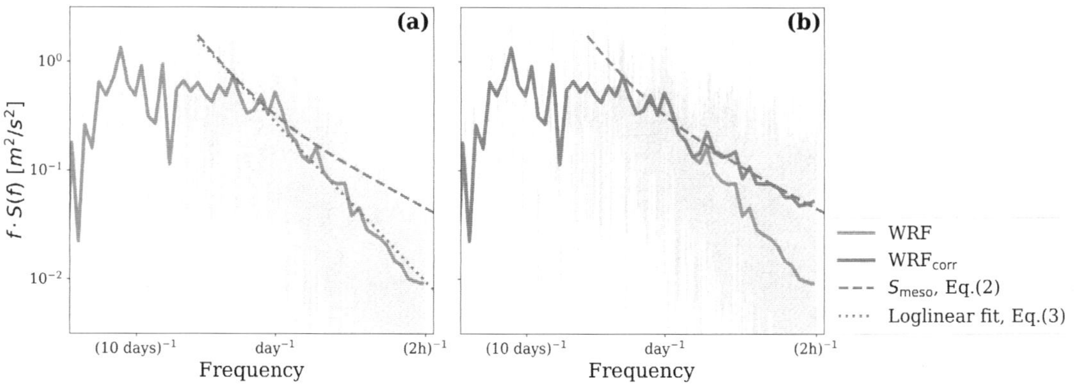

Figure 1. WRF spectra before (a) and after (b) spectral correction. The original WRF spectrum is shown in blue ("WRF") and the spectrum after correction in green ("WRF$_{\text{corr}}$"). To visualize the correction applied, the fitted spectral models of equations (2) and (3) are shown as dashed and dotted lines, respectively.

[20, 21]. The generation of the microscale wind fluctuations starts from a list of randomly generated complex numbers with mean 0 and unit variance. Then, the spectral values for each frequency, $S_{\text{meso+micro}}(f)$ from equation (6) are imposed to ensure that the Fourier transform will generate wind speed time series that are coherent with this spectrum.

Following the Nyquist-Shannon sampling theorem, the maximum resolved frequency by the mesoscale simulations is $f_{\text{Nyq,input}} = f_{\text{s}_{\text{input,WRF}}}/2 = 1/\left(2T_{\text{input,WRF}}\right)$. Therefore, high frequency spectral components need to be generated from $f_{\text{s}_{\text{input,WRF}}}/2$ to the desired output frequency $f_{\text{output}}/2$ to close the gap with the low-frequency mesoscale simulations. These stochastic simulations are then performed in blocks of length $2T_{\text{input, WRF}}$. For each block the average WRF wind speed, $\bar{U}_{\text{WRF}}$, and friction velocity, $\bar{u}_{*_{\text{WRF}}}$, is calculated as input for the spectral model of equation (6). For the friction velocity, $\bar{u}_{*_{\text{WRF}}}$ is first corrected to the microscale using:

$$u_{*_{\text{micro}}} = \bar{u}_{*\text{WRF}} \frac{\log\left(z/z_{0_{\text{WRF}}}\right)}{\log\left(z/z_{0_{\text{micro}}}\right)}, \tag{7}$$

where $z_{0_{\text{WRF}}}$ is the mesoscale roughness from the WRF simulations and the microscale roughness length, $z_{0_{\text{micro}}}$, can be estimated from roughness maps with the "equilibrium surface roughness" approach as defined in [1]. With the inclusion of $\bar{U}_{\text{WRF}}$ and $u_{*_{\text{micro}}}$ in the meso-micro spectral model, the non-stationarities of both quantities are inherently introduced in the microscale stochastic simulations. Finally, the combination of the different $2T_{\text{input,WRF}}$ blocks results in a simulated turbulence spectrum that covers the high frequency range in the frequency domain.

In this study with hourly WRF simulations, the spectrum will be simulated from $(2\,\text{h})^{-1}$ to $(2\,\text{s})^{-1}$ to arrive at wind time series with an output frequency of 1 Hz. An example of this simulated turbulent spectrum together with the meso-boundary-layer spectral model is shown in Figure 2. For representation purposes, the meso-micro spectral model uses the mean wind speed and friction velocity of the WRF simulations. As the spectral model used for microscale simulations is not constant but actually varies depending on the $\bar{U}_{\text{WRF}}$ and $u_{*_{\text{WRF}}}$ of each $2T_{\text{input,WRF}}$ block, there is not a perfect match between the simulated microscale spectrum and the spectral model as shown in Figure 2. Furthermore, the simulated spectrum in grey also shows that there is some remaining energy content in the lower frequency range ($f < (2\,\text{h})^{-1}$). This originates from the 2 h mean wind speed and friction velocity that drive the turbulence simulations and leak in some energy content from the mesoscale fluctuations.

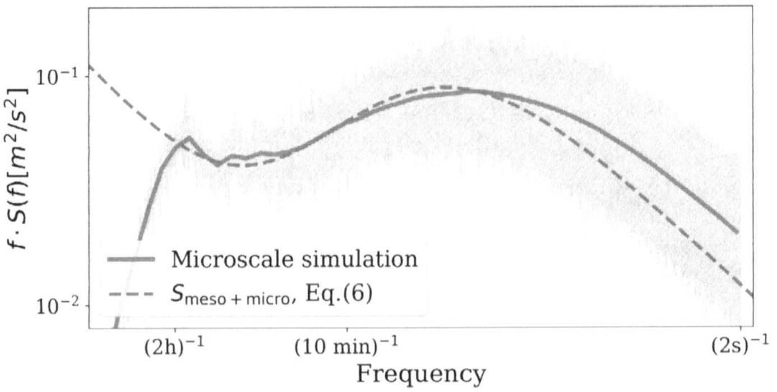

Figure 2. Simulated microscale spectrum to model the high-frequency turbulence fluctuations (grey) together with the meso-boundary-layer spectral model of equation (6) using the mean WRF wind speed and friction velocity (red dashed line).

3.3. Time series generation

For the time series generation, the corrected mesoscale spectrum from equation (4) is joined with the microscale turbulence spectrum from the stochastic simulations as described in subsection 3.2. Here, the corrected WRF simulations will reproduce the long-term non-stationary variations, while the microscale fluctuations in the high-frequency range are covered by the turbulence spectrum. This is visualised in Figure 3. The leakage from the WRF simulations into the microscale spectrum is first removed by a high-pass frequency filter before combining it with the corrected WRF spectrum to ensure a smooth spectrum across all scales.

As a final step, a Fast Fourier Transform (FFT) converts the simulated spectrum from the frequency into the time domain with a time resolution of dt. By grouping the high-resolution time series into 10-minute periods, realistic mean and standard deviation wind speed values can be obtained that resemble the typical output of a meteorological mast from a wind resource measurement campaign.

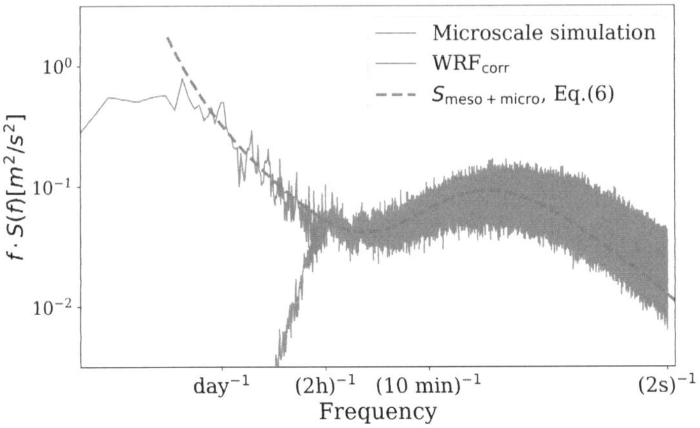

Figure 3. Combination of corrected low-frequency WRF spectrum representing the non-stationary weather patterns (green) with the high-frequency microscale spectrum from stochastic simulations (grey) to capture the turbulence variations. The full meso-micro spectrum used for the simulations is shown as a dashed red line.

4. Results and validation

The final combined $\mathrm{WRF_{corr+micro}}$ modelled spectrum is shown together with the measured spectrum in Figure 4. The validation shows a good agreement between the simulated and observed spectrum from the mast measurements throughout the entire spectral domain. This indicates that the methodology outlined in section 3 is able to generate wind data with physically realistic fluctuations in the mesoscale, gap (the so-called "terra incognita") and microscale regions that resemble actual field observations.

The simulated spectrum can be converted with a FFT to the time domain with a time resolution of 1 second before resampling to 10-minute mean and standard deviation values. Figure 5 shows the first 3 days of high-resolution simulated time series together with the original WRF wind speeds and the observations. The figure shows that the fluctuations of the simulated time series behave in a similar manner as the ones observed from the mast measurements.

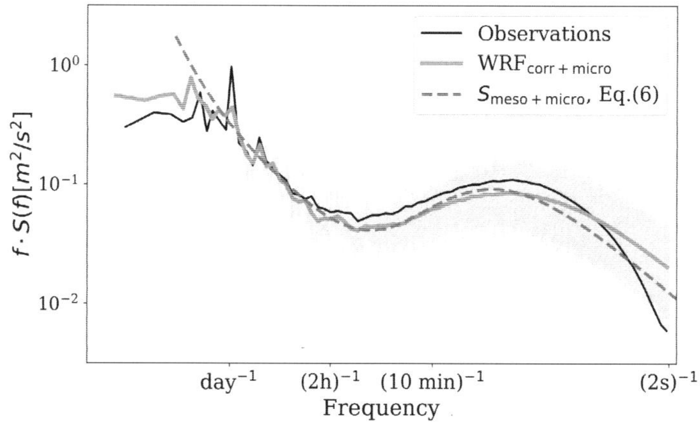

Figure 4. The superposition of the corrected WRF spectrum with the turbulence stochastic simulations (orange) is in good agreement with the observed spectrum (black) throughout the entire boundary-layer spectrum. As a reference, the spectral model used for the simulations is shown as a dashed red line.

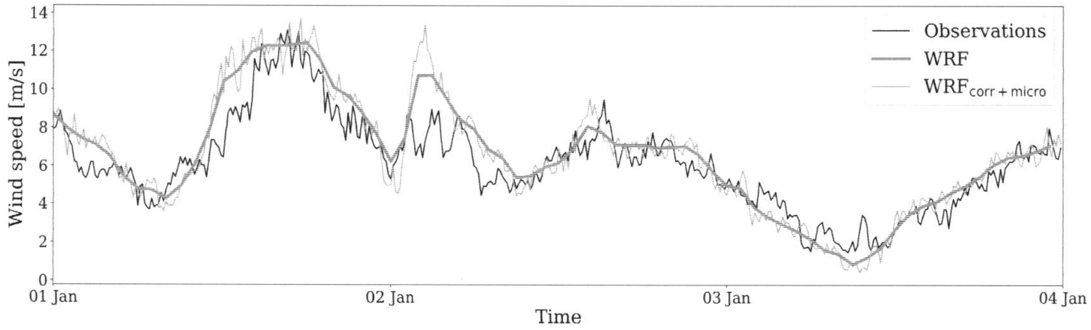

Figure 5. Example of wind speed time series from original hourly WRF simulations (blue), the WRF simulations temporally downscaled to 10 minutes (orange) and the mast measurements (black).

Finally, for wind power integration studies, the ramp behaviour is often investigated [14, 22]. For this, the wind speed ramp amplitude, ΔU, for a predefined time difference, Δt, is calculated:

$$\Delta U_{\Delta t} = U_t - U_{t-\Delta t}. \quad (8)$$

Then, probability and cumulative density functions (PDF and CDF, respectively) for a selected time window can be calculated to verify the ramp rate distribution. Figure 6 shows the PDF and CDF for the 10-minute wind speed differences for both the simulated as measured time series. Although somewhat fatter tails are observed in the measurements, overall, the modelled 10-minute ramps are closely related to the observed ramp distribution. The slight underestimation of the larger wind speed differences is likely to originate from the assumption of Gaussian distributed turbulence during the stochastic simulations. This deviation from the non-Gaussian nature of atmospheric turbulence can result in an underestimation of the wind turbine extreme and fatigue loads as well [3, 23]. However, as a turbine acts as a low-pass filter, using Gaussian or non-Gaussian turbulence inflow does not necessarily lead to significant differences in the mechanical loads that take time to develop [24]. The relevance of Gaussian versus non-Gaussian turbulence fields on turbine load and fatigue damage therefore still remains an ongoing research discussion.

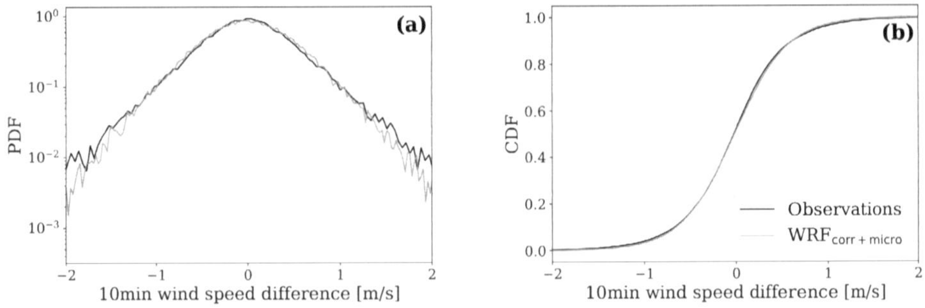

Figure 6. PDF (a) and CDF (b) of the 10-minute ramp rates for the WRF temporally downscaled simulation (orange) compared to the observations (black).

5. Conclusions and outlook

NWP models tend to produce wind speed time series that are excessively smooth and have a limited time resolution. This paper presents a new methodology to temporally downscale mesoscale simulations and generate realistic wind speed data at high temporal resolution without the need for CFD or on-site measurements. This method is demonstrated by coupling mesoscale simulations from the WRF model with stochastic turbulence simulations using a full-scale spectral model to downscale the hourly wind speeds down to 10-minute mean and standard deviation values.

This hybrid approach to couple low-frequency mesoscale time series with high-frequency turbulence simulations can be summarized as follows:

(i) Correct for the smoothing of mesoscale simulations.
   (a) Compute the PSD of the numerical simulations.
   (b) Calibrate the Larsén mesoscale spectrum to observed PSD.
   (c) Correct spectral components to the Larsén mesoscale spectrum model for higher frequencies.

(ii) Extend the mesoscale spectrum to the microscale range.
   (a) Generate microscale spectral components to any desired frequency $1/dt$ using a full-scale boundary layer spectral model.
   (b) Combine the low frequency corrected mesoscale spectrum (i) with the high frequency microscale components (iia).

(iii) Generate time series from the combined modelled spectrum.
   (a) Transform the combined spectrum to time domain with a FFT.
   (b) Resample the high frequency time series to 10-minute mean and standard deviation values to resemble field observations.

The resulting wind speed time series correctly represents non-stationary weather patterns introduced by the mesoscale simulations, as well as the high-frequency turbulent motions captured by the boundary layer spectral models. Using these realistic time series data in wind studies can typically lead to more accurate production and loss estimations. Furthermore, it allows wind experts to improve their assessments of ramping and extreme events or assist in the dimensioning of storage systems.

A more extensive validation is planned to verify the performance of the temporal downscaling methodology at multiple sites with different topographical and atmospheric conditions. Furthermore, the modelling approach can be extended by including the downscaling of the lateral velocity component, v, to reproduce high frequency wind direction variations as well. This can be achieved with the same stochastic simulation principles as discussed in subsection 3.2 together with the latest advancements in the full-scale v spectral model from [21].

Acknowledgements

The work on this paper is part of the WindSider project. It has been carried out with the support of the EU Horizon 2020 call H2020-EIC-FTI-2018-2020 (Agreement n° 878788).

References

[1] Troen I and Lundtang Petersen E 1989 *European Wind Atlas* (Risø National Laboratory) ISBN 87-550-1482-8
[2] Jourdier B and Drobinski P 2017 *Annales Geophysicae* **35** 691–700
[3] Wächter M, Heißelmann H, Hölling M, Morales A, Milan P, Mücke T, Peinke J, Reinke N and Rinn P 2012 *Journal of Turbulence* N26
[4] Deane J, Drayton G and Gallachóir B Ó 2014 *Applied Energy* **113** 152–8
[5] Troy N, Flynn D and O'Malley M 2012 The importance of sub-hourly modeling with a high penetration of wind generation *2012 IEEE Power and Energy Society General Meeting* (IEEE) pp 1–6
[6] Al-Yahyai S, Charabi Y and Gastli A 2010 *Renewable and Sustainable Energy Reviews* **14** 3192–8
[7] Staffell I and Pfenninger S 2016 *Energy* **114** 1224–39
[8] Hahmann A N, Sīle T, Witha B, Davis N N, Dörenkämper M, Ezber Y, García-Bustamante E, González-Rouco J F, Navarro J, Olsen B T *et al.* 2020 *Geoscientific model development* **13** 5053–78
[9] Bastine D, Larsén X, Witha B, Dörenkämper M and Gottschall J 2018 *Journal of Physics: Conference Series* **1102** 012006 URL `https://doi.org/10.1088/1742-6596/1102/1/012006`
[10] Skamarock W C 2004 *Monthly weather review* **132** 3019–32
[11] Larsén X G, Ott S, Badger J, Hahmann A N and Mann J 2012 *Journal of Applied Meteorology and Climatology* **51** 521–33
[12] Rodrigo J S, Arroyo R C, Gancarski P, Guillén F B, Avila M, Barcons J, Folch A, Cavar D, Allaerts D, Meyers J and Dutrieux A 2018 *Journal of Physics: Conference Series* **1037** 072030 URL `https://doi.org/10.1088/1742-6596/1037/7/072030`
[13] Dörenkämper M, Olsen B T, Witha B, Hahmann A N, Davis N N, Barcons J, Ezber Y, García-Bustamante E, González-Rouco J F, Navarro J *et al.* 2020 *Geoscientific model development* **13** 5079–102
[14] Koivisto M, Jónsdóttir G M, Sørensen P, Plakas K and Cutululis N 2020 *Renewable Energy* **159** 991–9
[15] Larsén X G, Larsen S E and Petersen E L 2016 *Boundary-layer meteorology* **159** 349–71
[16] Larsén X G, Larsen S E, Petersen E L and Mikkelsen T K 2019 *Boundary-Layer Meteorology* **171** 191–212
[17] Rose S and Apt J 2012 *Wind Energy* **15** 699–715
[18] Larsén X G, Vincent C and Larsen S 2013 *Quarterly Journal of the Royal Meteorological Society* **139** 685–700
[19] Mikkelsen T, Larsen S E, Jørgensen H E, Astrup P and Larsén X G 2017 *Physica Scripta* **92** 124002
[20] McFarlane A, Veers P and Schluter L 1994 Simulating high-frequency wind for long durations *Thirteenth ASME Wind Energy Symposium* (New Orleans, LA)
[21] Larsén X G, Larsen S E, Petersen E L and Mikkelsen T K 2021 *Boundary-Layer Meteorology* **178** 415–34
[22] Olauson J, Bergkvist M and Rydén J 2017 *Wind Energy* **20** 973–85
[23] Schwarz C M, Ehrich S and Peinke J 2019 *Wind Energy Science* **4** 581–594 URL `https://wes.copernicus.org/articles/4/581/2019/`
[24] Berg J, Natarajan A, Mann J and Patton E G 2016 *Wind Energy* **19** 1975–89

The influence of atmospheric stability on wind turbine performance: a quantitative study using measurement from SCADA system and mast

Minh-Thang DO, Ru LI, Zixiao JIANG

minh-thang.do@meteodyn.com

Abstract. In the wind resource assessment, to estimate the energy potential of a site, engineers use the manufacturer power curve. This power curve is an ideal relationship between the wind speed of an ideal wind flow and the output power of the designed turbine. The industry standard IEC 61400-12-1 establishes some requirements to measure such power curves. In the real condition, with the same value of wind speed, a large range of output power could be obtained. This difference may be caused by atmospheric stability, which affects wind profiles across the turbine blades. This study focuses on the influence of atmospheric stability on wind turbine performance in real condition. It is shown that the atmospheric stability calculated using the standard wind measurements from a conventional weather mast could be applied for every turbine in a medium size wind farm (33 wind turbines). This influence is very sensitive to the shape factor k of the wind speed distribution. The loss of AEP between neutral/stable conditions could be neglected while the loss of AEP between unstable/stable conditions is significant and depends on the average wind speed. In practice, it is possible to consider this influence during the wind resource assessment study while knowing the percentage of each atmospheric stability condition and the distribution of wind speed on the site.

1. Introduction

In the wind resource assessment, to estimate the energy potential of a site, engineers use the manufacturer power curve. This power curve is an ideal relationship between the wind speed of an ideal wind flow and the output power of the designed turbine. The industry standard IEC 61400-12-1 [1] establishes some requirements to measure such power curves.

In the real condition, with the same value of wind speed, a large range of output power could be obtained. This difference may be caused by atmospheric stability, which affects wind profiles across the turbine blades.

The impact of atmospheric stability to the wind turbine performance has been studied in several research [2], [3]. However, these studies require a sonic detection and ranging instrument (SODAR) which is expensive and not available in popular weather mast. Moreover, they focus mostly on the calculation of the atmospheric stability and do not provide a detailed analysis for the losses of production due to the atmospheric stability.

Therefore, based on a novel method to quantify atmospheric stability from standard wind measurements [4] and our experience in analysing the wind turbine performance, the influence of atmospheric stability on wind turbine performance will be analysed in a quantitative study using measurements from the scada system and conventional weather mast.

2. Approach

2.1. Site and data description

The DiQingYuan wind farm considered in this study is consisted of 33 wind turbines in which:

- 29 turbines are model E93 with a nominal power of 1.5MW, rotor diameter of 93m
- 4 turbines (D9 D25 D26 D31) are model E103 with a nominal power of 1.6MW, rotor diameter of 103m

The SCADA system collects the data of wind speed and active power at a frequency of 5 minutes for each turbine. The SCADA data is available from 01/01/2017 to 05/03/2018.

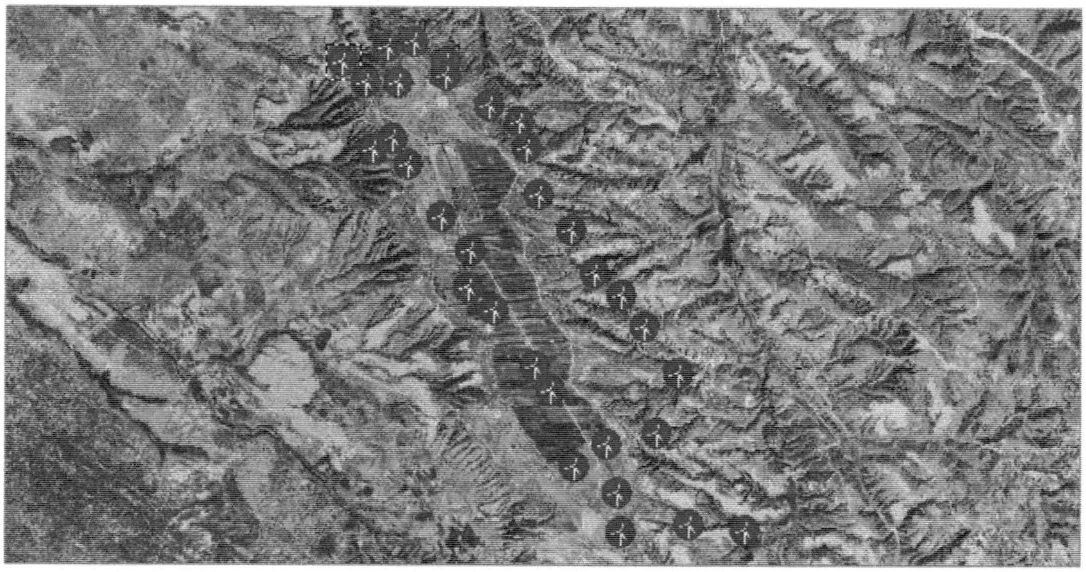

Figure 1. Satellite view of the wind farm

A weather mast is installed in the middle of the wind farm, which measures the wind at 5 different heights: 10m, 30m, 50m, 70m and 80m. For each height, 3 variables are collected: wind speed, standard deviation, and wind direction. The data frequency of the data collected by the weather mast is also 5 minutes. The mast data is available from 01/01/2017 to 31/12/2019.

In total, only data from 01/01/2017 to 05/03/2018 is usable. After data cleansing and formatting, the amount of valid data in this study is 422 days, which is adequate for a solid study.

2.2. Atmospheric stability calculation and classification

To study the atmospheric stability, two indicators, the Monin-Obukhov length (MOL) or Richardson number (Ri), are used to estimate the effects of buoyancy on turbulent flow. The general method to quantify the effects of buoyancy from the wind measurements, calculates the MOL through the heat flux or the temperature gradient, which are not provided in the standard wind measurements.

The wind shear is often selected to quantify the atmospheric stability. But during the unstable conditions, the power law is not an accurate approximation for the wind speed profile [5]. The relationship between TKE and the atmospheric stability is not proved [6].

Svenningsen [4] suggests quantifying the atmospheric stability through the Turbulent Kinetic Energy (TKE) range and wind shear. In his method, it is needed to estimate the neutral wind shears ($\widehat{\alpha}(\theta)$) and the turbulence intensity ($\widehat{TI}(\theta)$) at sector θ from the highest 1-3% of wind speeds. Then, we calculate the deviation of the wind shear α and the turbulence intensity TI,

$$\delta TI(t) = \frac{TI(t)}{\widehat{TI}(\theta)} - 1, \delta\alpha(t) = \frac{\alpha(t)}{\widehat{\alpha}(\theta)} - 1$$

We distinguish the atmospheric stability situations by the sign of the deviations:

$$\text{Neutral: } \sqrt{(\delta TI(t))^2 + (\delta\alpha(t))^2} \to 0$$
$$\text{Stable: } \delta TI(t) < 0 \ \& \ \delta\alpha(t) > 0$$
$$\text{Unstable: } \delta TI(t) > 0 \ \& \ \delta\alpha(t) < 0$$

Svenningsen's method is calibrated through the measurements in Cabauw. Before using his method to analyze the power in the wind farm DiQingYuan, we did a second validation with the measurements in Rödeser Berg in German, where we have the measured MOL. We find that the Svenningsen's method obtains an equivalent atmospheric stability situation comparing with the one obtained by measured MOL.

3. Analysing the influence of atmospheric stability on wind turbine performance

Using the classification of atmospheric stability, we could classify the datapoints from SCADA system into 3 categories: unstable (173 days – 41%), neutral (128 days – 30%), and stable (121 days – 29%). The influence of atmospheric stability on wind turbine performance is then be analysed by comparing the wind turbine performance in these 3 categories.

3.1. Influence on the power curve

The loss of performance in unstable condition compared to stable condition is observed at every wind turbine in the wind farm through a significant gap between the two power curves. This statement is not only valid with the wind turbine which is close to the mast (D07) but also with the farthest wind turbine (D32). It means that the atmospheric stability calculated from the mast is valid for the whole wind farm at a SCADA data frequency of 5 minutes. It is also noticed that the gap between power curves in neutral condition and stable condition is relatively small, which means the loss of performance is limited.

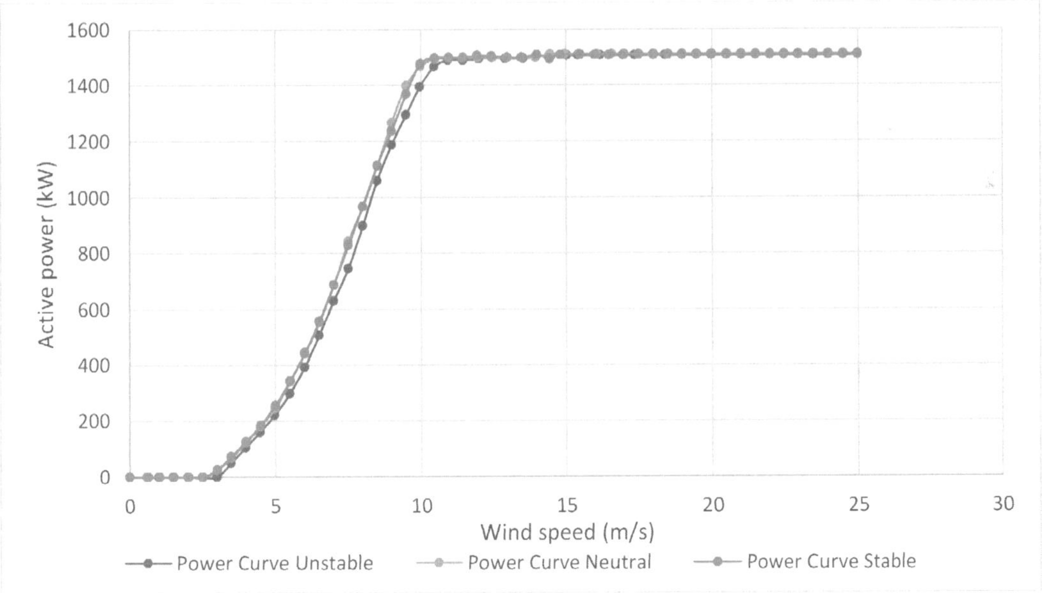

Figure 2. Power Curve of Wind turbine D07 in different stabilities

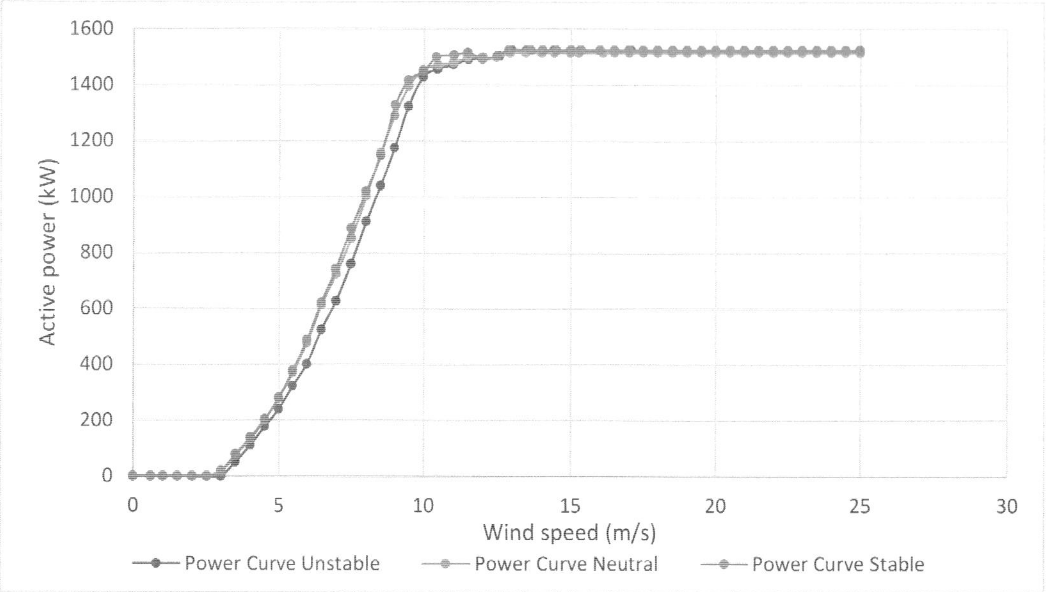

Figure 3. Power Curve of Wind turbine D32 in different stabilities

However, it is quite difficult to have a quantitative comparison based on the power curve. Therefore, a statistics indicator, the Annual Energy Production (AEP), is used.

3.2. Influence on the Annual Energy Production

The AEP of each wind turbine will be calculated using the same wind distribution for the whole wind farm in each atmospheric stability condition. From the wind speed collected at 33 nacelles for 422 days, the coefficients of the Weibull distribution are calculated:

- Scale parameter A = 5.58 (m/)
- Shape parameter k = 2.13

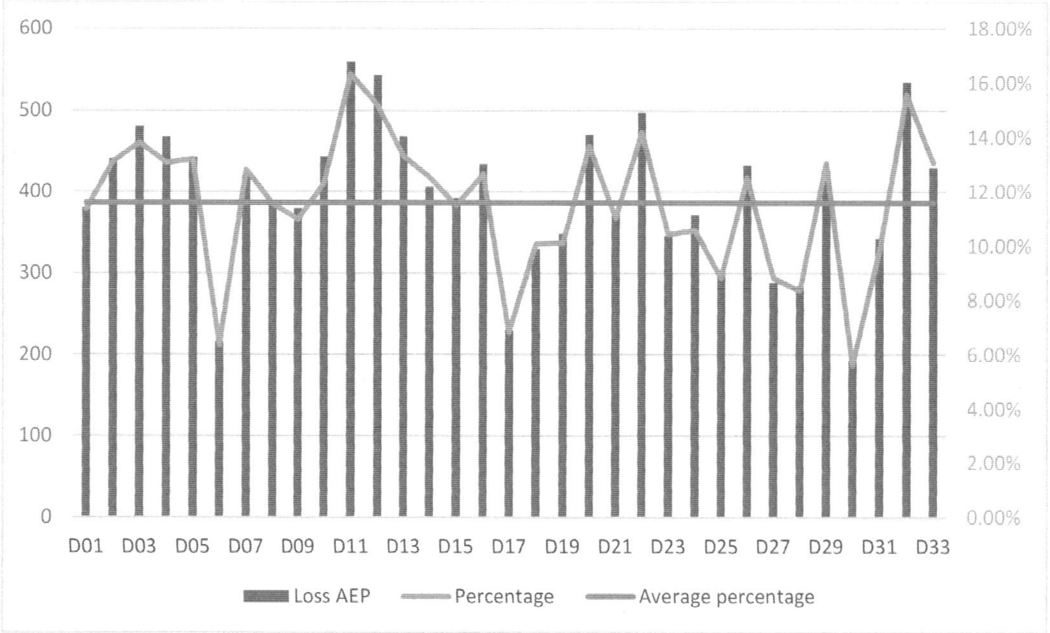

Figure 4. Loss of AEP in each wind turbine between unstable and stable

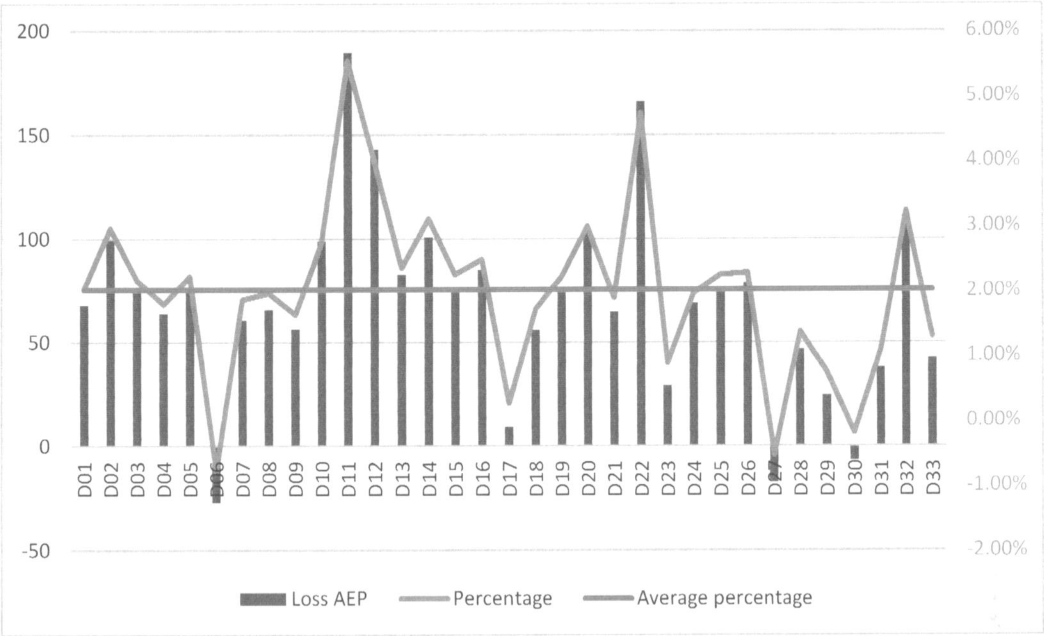

Figure 5. Loss of AEP in each wind turbine between neutral and stable

It is obvious that the AEP in stable condition is the highest, followed by the neutral condition. The AEP in unstable condition is the worst. The loss of AEP in each wind turbine between unstable and stable and between neutral and stable conditions is presented in Figure 4 and Figure 5. The average loss of the wind farm between unstable and stable conditions is 11.59% and between neutral and stable conditions is only 2.02%.

3.3. Annual Energy Production following IEC 61400-12-1

To facilitate the comparison of the AEP, IEC propose to calculate the AEP of wind turbine using a Rayleigh distribution, which is identical to a Weibull distribution with a shape factor of 2 and with annual average wind speeds of 4 m/s, 5 m/s, 6 m/s, 7 m/s, 8 m/s, 9 m/s, 10 m/s and 11 m/s.

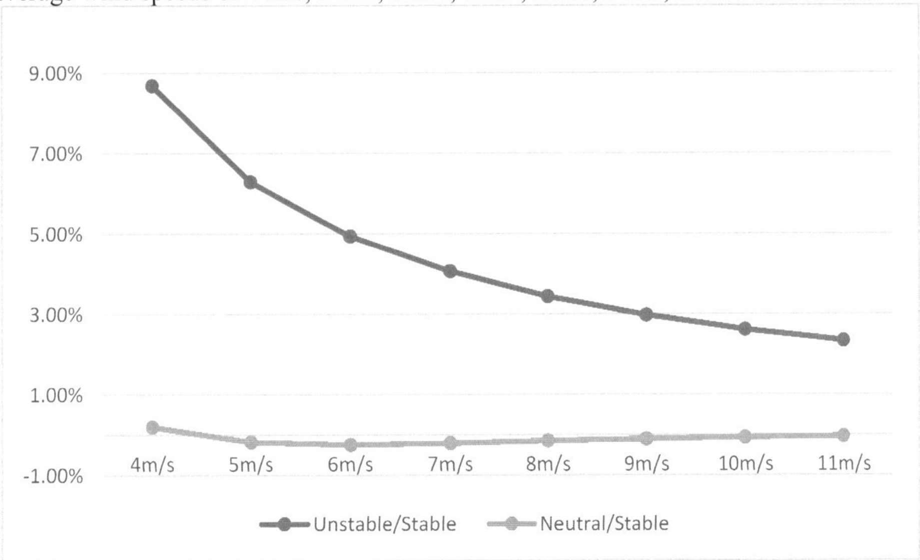

Figure 6. Loss of AEP between unstable/stable and between neutral/stable for each level of annual average wind speed

The average loss of the wind farm between unstable and stable condition and between neutral and stable condition is then calculated for each level of annual average wind speed.

It could be seen that the loss of AEP between neutral/stable conditions is close to zeros and then could be neglected while the loss of AEP between unstable/stable conditions is significant and depends on the average wind speed. Starting from 8.69% at the average wind speed of 4 m/s, this loss decreases to 2.35% at 11m/s (Figure 6). This should be and could be considered in the wind resource assessment study.

The loss presented in this section is lower than 11.59% from the previous section (or 2.02%) because there's a small difference in the shape factor k of the distribution (2 instead of 2.13), which shows that this loss is very sensitive with the shape factor.

4. Conclusions

This study focuses on the influence of atmospheric stability on wind turbine performance in real condition. It is shown that the atmospheric stability calculated using the standard wind measurements from a conventional weather mast could be applied for every turbine in a medium size wind farm (33 wind turbines).

This influence is very sensitive to the shape factor k of the wind speed distribution.

The loss of AEP between neutral/stable conditions could be neglected while the loss of AEP between unstable/stable conditions is significant and depends on the average wind speed.

In practice, it is possible to consider this influence during the wind resource assessment study while knowing the percentage of each atmospheric stability condition and the distribution of wind speed on the site.

References

[1] IEC 61400-12-1, «Wind turbines – Part 12-1: power performance measurements of electricity producing wind turbines,» IEC – International Electrotechnical Commission, Geneva, Switzerland, 2005.

[2] S. Wharton et J. K.Lundquist, «Atmospheric stability affects wind turbine power collection,» *Environmental Research Letters,* vol. 7, 2012.

[3] X. Han, D. Liu, C. Xu et W. Z. Shen, «Atmospheric stability and topography effects on wind turbine performance and wake properties in complex terrain,» *Renewable Energy,* vol. 126, pp. 640-651, 2018.

[4] L. Svenningsen, R. M. M. Slot et M. L. Thøgersen, «A novel method to quantify atmospheric stability,» *Journal of Physics: Conference Series,* vol. 1102, 2018.

[5] S.Emeis, «A simple analytical wind park model considering atmospheric stability,» *Wind Energy,* 2010.

[6] C. L. Archer, B. A. Colle, D. L. Veron, F. Veron et M. J. Sienkiewicz, «On the predominance of unstable atmospheric conditions in the marine boundary layer offshore of the U.S. northeastern coast,» *Journal of Geophysical Research: Atmospheres,* 2016.

Comparison of novel SCADA Data Cleaning Technique for Wind Turbine Electric Pitch System

C McKinnon [1], K Tartt [1], J Carroll [1], A McDonald [2], C Plumley [3] and D Ferguson [4]

University of Strathclyde, 99 George Street, Glasgow, G1 1RD

E-mail: `conor.mckinnon@strath.ac.uk`

Abstract. Wind turbines typically do not operate in the ideal operating conditions, leading to abnormal behaviour that is reflected in their power curves. This abnormal behaviour can affect the performance of condition monitoring processes, as it may mask faulty behaviour. By cleaning other abnormal data, such as curtailment, models can learn the normal behaviour of the turbines.

This paper presents a novel cleaning technique that utilises a combination of data binning and the Mahalanobis distance. This removes between 5 to 6% of the data, without great loss of normal data. When compared against other data cleaning techniques, the one presented in this paper produces a more ideal power curve. This technique could improve the performance of data-based condition monitoring techniques.

1. Introduction

In the decade from 2009 to 2019, there has been an eleven-fold increase in offshore wind power generation in Europe, illustrated in figure 1, this has taken offshore wind generation from 2 GW of generation to 22 GW. There has also been over 100 GW of increased generation onshore at the same time[1]. This has been a substantial change to the energy mix of the EU, and there are even greater increases set to come in the next decade.

There is a massive drive to install more wind capacity in the EU, and this will lead to even larger farms of turbines in operation requiring greater oversight. Operators will also be overseeing a greater number of turbines than ever before. With this greater capacity, operations and maintenance (O&M) will require better techniques and strategies to improve efficiency and reduce the costs. According to Wood MacKenzie[2], O&M costs were projected to be $ 15 Billion in 2019, and 57% of this would be on unplanned repairs and corrective maintenance.

Wind farms offshore have a failure rate per year on average of 10 per turbine. Of this, 20% of failures are major maintenance actions. The pitch system has one of the highest failure rates, and one of the lowest repair times[3], however offshore a high failure rate can lead to exacerbated downtimes even if the repair time is low. This is due to the lower accessibility offshore, which can be accounted for by weather conditions and vessel availability. According to [4] the pitch system, which was considered part of the hub, had an average downtime of 0.64 days per year, and another paper[5] has stated that the system fails between 0.1 and 0.3 times per year.

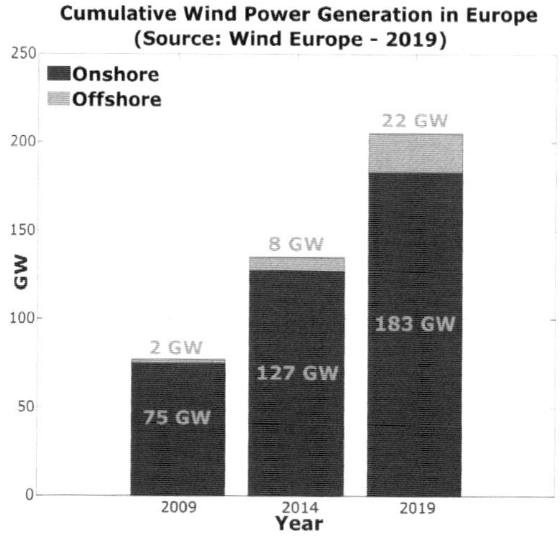

Figure 1. Illustration of increase in Offshore wind in Europe, from 2009 to 2019. Data taken from Wind Europe[1].

Turbine downtime can be reduced by utilising condition-based maintenance, over conventional methods such as corrective or scheduled maintenance. This approach can reduce the frequency of unplanned downtimes, and the associated costs. This requires both monitoring and analysing the component condition, and then scheduling maintenance based on the condition[6] - allowing to plan around weather and vessel availability. One method of analysis that is being considered is normal behaviour modelling, which tries to build a model of the normal data and predict future performance based on that - then if there is an error this can be analysed for potential warning signs.

The performance of condition monitoring techniques can be improved by using data pre-processing before the analysis. One such method is data-cleaning, where data outliers are removed from the data-set to improve the performance of normal behaviour modelling. In this paper a novel data cleaning technique is presented based on the use of the Mahalanobis distance. The data was cleaned through binning of wind turbine power curves, then applying Mahalanobis distance to each bin. This method was also compared against several other techniques for validation.

This paper aims to provide a novel user friendly data cleaning technique to remove anomalous data without much loss of information. The objectives of this paper is to test the presented technique on multiple turbines, and against other similar techniques. This testing is done on various metrics, to help assess the performance of the techniques examined. To the best of authors' knowledge, the use of Mahalanobis distance with data binning has not been used previously for cleaning of wind turbine power curves. The use of this in combination with an above rated filter has also previously not been considered.

The structure of the paper is as follows. Section 2 reviews previous literature in condition monitoring for wind turbines. Section 3 outlines the method used in this paper along with a description of the data in the case studies considered. Finally, section 4 presents and discusses the results from the method outlined in section 3.

2. Literature Review

Several papers have presented novel techniques for wind turbine data cleaning in recent years. One example of a novel data cleaning technique was produced by Xiaojun Shen et al.[7]. The authors used change point detection (CPD) on the variance of the binned power values taken from SCADA data. This was the first step of data removal, and then a second step using 1.5 times the inter-quartile range (IQR) to remove outliers above the wind power curve, as CPD could not remove these effectively. The data deletion rate was used as a comparison metric, and proved to perform better than Local Outlier Factor (LOF) and IQR then CPD, instead of CPD then IQR.

A paper from A. Li Yuan et al.[8] used least squares to model a fitted power curve from SCADA data. This was fit to the average wind speed - power values from data bins. Upper and lower control limits, consisting of 3 standard deviations from the best fit line, were used to remove abnormal data.

Image based techniques have been used previously by Y. Su et al,[9]. The power curve was converted to a binary image with any pixel containing data converted to white. The data was considered anomalous based on their median distance to the bounday of a bright pixel group.

Another paper[10] has looked at the use of K-means clustering to initialise cluster centres before then clustering again with manifold spectral clustering.

Another image based method, from H. Long et al.[11], converted scatter data to a binary image. The principal part of the power curve is extracted and the Hu moments calculated. The extracted reference curve is considered normal, and anything outside this boundary is abnormal.

A paper by J. Wu[12] modelled a power curve by finding wind speed values and calculate the corresponding power values. Centre values of these values were used to plot the power curve, and upper and lower envelope lines obtained. This removed much of the outliers, then the data was binned and the Inter Quartile range was applied to the data to remove the rest of the outliers.

A similar technique from Shen, Fu and Zhou[7], first applied Change Point Detection to binned and ordered power values to first find the lower change point. This removed much of the outliers in the power curve. However, outliers still occurred above and below the curve, so the Inter Quartile Range was used to remove the rest.

Another image based method, from Z. Wang et al.[13], converted the curve to a binary image. First non-continuous vertical pixels were removed, then this was repeated for the horizontal direction. This removed much of the outliers in the curve.

A paper by Z. Lin et al.[14] compared the use of elliptic envelope and isolation forest to cleaned power curves from the Levenmouth 7 MW demonstrator turbine. This was done to pre-process data before use in a predictive deep learning neural network. Isolation forest was found to be better than Elliptic Envelope for this purpose.

A paper from Y. Bao et al.[15] utilised k-nearest neighbours in data bins to remove outliers before approximating the power cuvre with a least squares B-spline curve.

Another paper that utilises the inter quartile range (IQR) is from T. Yuan[16], which first clusters data using DBSCAN into core, border, and noise clusters. The noise is first removed, then the IQR is used to set upper and lower data limits to remove the remainder of the outliers.

3. Methodology

3.1. Data

The data provided for this paper was taken from the Supervisory Control and Data Acquisition (SCADA) system of four multi-megawatt wind turbines situated in South America. The SCADA system records data for multiple wind turbine components and down-samples it to 1 recording every 10 minutes. Each data record was approximately 2 years in length, and each was split evenly into training and testing datasets.

Two of the turbines provided did not fail within the recording period, these are classed as healthy turbines in this paper. The other two turbines did fail within the recording period, and these are classed as unhealthy. The unhealthy turbines failed from the same fault in the pitch system, which was located in the bearing. The healthy turbines were used to assess the technique's ability to retain important information when removing data. The unhealthy turbines were used to assess how well the overall method worked with those turbines that failed during the time period examined.

3.2. Data Cleaning

Several techniques were compared in this paper, and each consisted of multiple stages. The first stage in the model was to either remove, any data with either a negative power value, or a power value over 2500 kW (for simplicity this is referenced to as the negative power filter). These outliers are typically due to sensor error or turbine downtime. This stage was accomplished through simple filtering of the data in Python, and the effect can be seen in figure 2

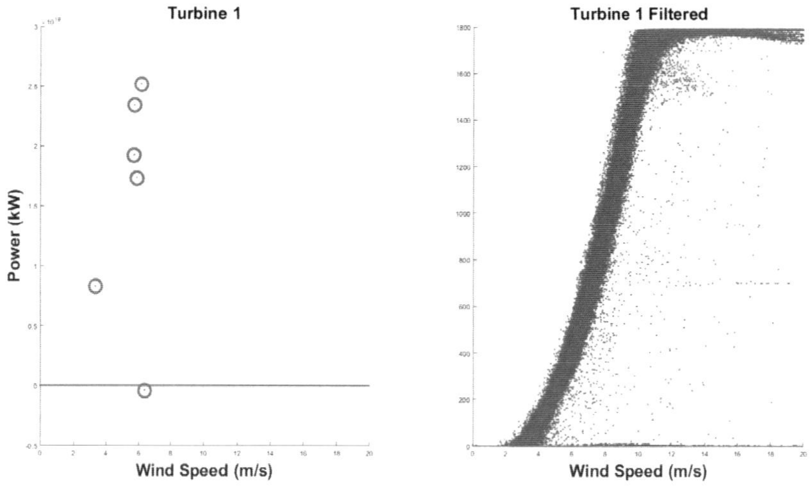

Figure 2. Comparison of Turbine 1 power curve, with and without filter. The red circles highlight where excessive readings occurred, and the orange line is positioned at 0 kW.

Second, data bins of roughly 1 m/s width were constructed. The mean power and wind speed values of these bins were found as cluster centres. The cluster centres found during binning were then used to find the Mahalanobis distance of each data-point in the bin. This is shown in equation 1:

$$D_{Mahalanobis}(x, y) = \sqrt{(x-y)C^{-1}(x-y)} \tag{1}$$

Mahalanobis distance appears to be more suitable for this data as it takes into account the covariance of the data. This is superior to the more conventional Euclidean distance, shown in equation 2, which assumes the data is distributed in a circle, whereas data in the power curve bins are distributed in truncated ellipses. The Euclidean distance was also compared in this paper, with the previous filtering and binning steps being the same. To remove data, a threshold of 2 standard deviations of the distance was used for both the Mahalanobis and Euclidean distances. This, in theory, should remove only around 5% of the data.

$$D_{Euclidean}(x, y) = \sqrt{(x_1 - x_2)^2 + (y_1 - y_2)^2} \tag{2}$$

Similarly, Local Outlier Factor (LOF) was compared using the same filtering and binning steps. This technique, described in detail in [17] first finds the k nearest neighbours to each point in the data, then the "reachability distance" of this point to others in the dataset. The average of these distances are found for the k nearest neighbours, which is called the local reachability distance. The LOF is then the average ratio between the local reachability distance of the point and its nearest neighbours.

Wind turbine blades are typically pitched to feather to control power output in wind turbines, this is typically done above rated wind speeds however during regimes of curtailment this can be done below rated. This can lead to straight horizontal lines under the power curve where the turbine is operating at a fixed power over a range of wind speeds. As another method for comparison, any datapoint corresponding to below rated pitch activity was removed. As this is one of the simpler techniques used currently, it was added as a baseline test to compare the slightly more complex methods against a relatively simple one. Here no binning was performed, however the negative power filter was still applied.

Finally, a comparison was made between the techniques described, and their counterparts that were only applied to data below rated. For the above rated data, a general filter of any power value lower than 1600 kW was removed. It was found that above rated the data was sparse, and it may not be suitable to use Euclidean or Mahalanobis distance as the standard deviation threshold assumes the data in a Gaussian Distribution. The power values above rated were scattered sparsely from 0 to over 2000 kW, and with most being clustered around rated power. The distribution of this is closer to a Poisson distribution, and therefore a filter to remove the tail of the data was considered. This change in distribution is brought on by the rated power acting as an upper bound on the data.

These techniques were compared against a technique from a paper by Shen, Fu, and Zhou[7] presented in Section 2. This involved the use of Change Point Detection (CPD) and Inter-Quartile Range (IQR). Binning was applied to the data, then data was initially removed using Change Point Detection, then Inter-Quartile Range was applied to the data to remove further outliers. This technique was compared against to provide context of this paper within the literature.

4. Results

4.1. Effectiveness of Technique

This section examines the effectiveness of the presented technique. The results for four turbines are compared, two of which did not fail in the time recorded and two did fail. For comparison, each turbine dataset was cleaned with and without the above rated filter presented in Section 3. Figure 3 presents the power curves for each of the turbines, here turbines 1 and 2 are considered healthy, and turbines 3 and 4 are unhealthy. The raw datapoints are plotted in blue, whilst the cleaned datapoints are in orange. The graphs on the right of the image also applied the above rated filter. The negative power filter was applied before the technique presented in this section.

Table 1 presents the percentage of data removed from the original datasets for each turbine by each cleaning technique, and the percentage of data removed from the filtered data. All models presented were applied on filtered data, except for the CPD-IQR method which was applied only on the raw data. The negative power filter appears to remove between 10 and 19% of data from the original, along with the outliers presented in figure 2. The models with the above rated filters applied are noted with an "AR" in table 1.

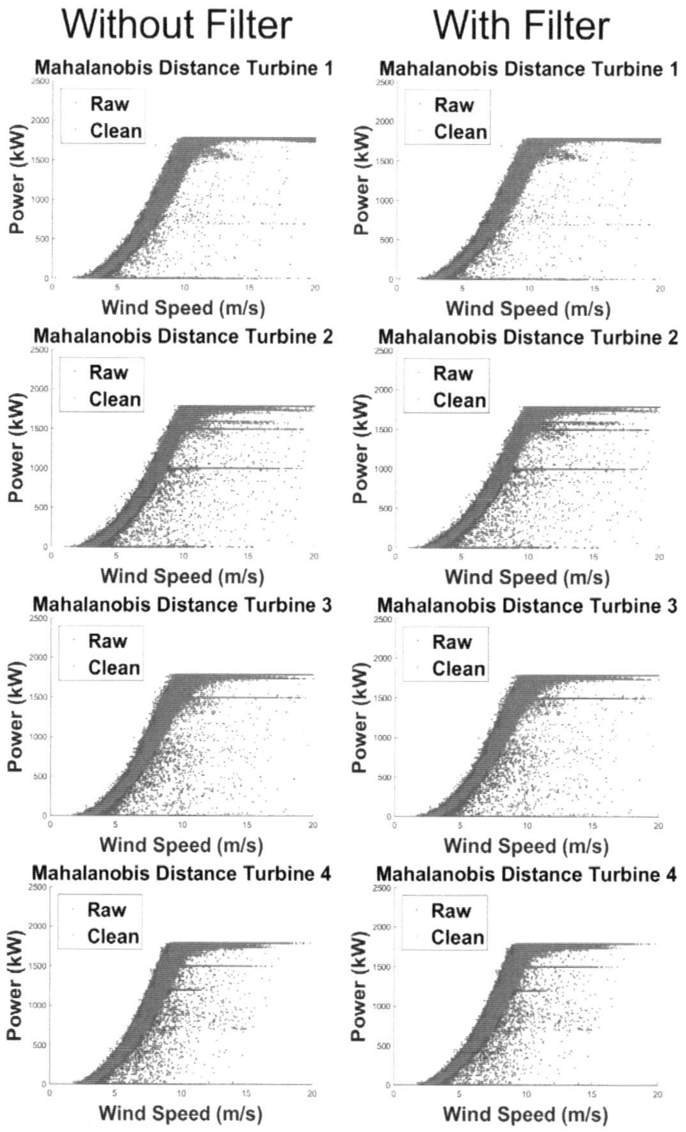

Figure 3. These graphs compare the effect of the cleaning technique with and without the above rated filter.

The effect of the above rated filter can be seen in figure 3, where more anomalies below the graph are removed. In particular for turbines 2, 3, and 4 the horizontal lines just below rated power, that typically indicate power curtailment, are filtered out. These were missed by the original cleaning model, and can affect normal behaviour models. By comparing the effect of the Mahalanobis model, with and without the above rated filter, it can be seen from table 1 that less than 1% of the data is removed by this filter. Wind turbines above rated power are controlled with the pitch system to remain at rated power at any wind speed, so any power values that vary from above rated can be considered different from normal wind turbine behaviour. Whilst wind power curtailment is planned behaviour by turbine operators, it is abnormal to the ideal power curve which aims to produce rated power at all times above rated wind speed, and below cut-out. Therefore, it is important to remove this data as it can affect the performance of power prediction, or fault detection, as this planned behaviour will affect the model and produce errors during normal turbine operation at above rated wind speeds. While most of the above rated outliers are removed by the Mahalanobis cleaning method, some curtailment data remains, therefore the use of the above rated filter here is appropriate in removing the extra data that can affect normal behaviour modelling. It is possible that by using the median wind and power values to determine bin centres that this could remove the need for the above rated filter. The mean of the power values will be pulled down by sparse low power values, whereas a median would remain within the more dense cluster near rated.

The model is shown to have similar effects on each turbine, removing much of the outlier data below the curve, and the outliers found above the curve, such as those in figure 2. This is evident for the models with and without the above rated filter. Of the four turbines, only turbine 1 appears to have very little curtailment above rated, compared to the others. For all turbines, it appears the Mahalanobis method removes between 5 and 6% of the negative power filter data, and the same method with the above rated filter removes between 5 and 7.5%. As the threshold used is based upon the standard deviation this is to be expected.

By looking at figure 3, the amount of normal information being retained by each model can be compared for each turbine. Each turbine appears to have most of the normal data retained above rated, with some normal data being removed from below rated. There is also a slight step effect in the below rated regime due to the binning of the data and applying the Mahalanobis distance method. This could potentially be solved by slightly overlapping the bins and only removing data where outliers are detected in both bins. Below rated there seems to be data removed from the main cluster of the power curve, it may be that the threshold is too low for abnormal data below rated, and therefore a threshold of 3 standard deviations could maybe be applied in the future below rated. This would need to be tuned to retain as much of the normal data as possible.

A way to perhaps improve on this technique is to introduce a level of supervision into the technique. This could be done by including data labels, such as curtailment, turbine faults, and normal data, even as a method of evaluating the technique. This context would help to identify data for removal, however only some of this data could be removed when applied online as fault data would not be available in real time. This could potentially allow for a first step of removing any data flagged as curtailment, then further data cleaning could be done, any data left that is abnormal would be sparse and random in nature which is well suited to this technique.

Overall, this technique appears to remove much of the abnormal data, without sacrificing too much normal data to do so. The above rated filter works well at removing curtailment data that is missed by the unfiltered technique. The ideal power curve is preserved by the Mahalanobis distance method, however there are some areas that could be improved by future work, such as the loss of some normal data below rated.

4.2. Comparison of Cleaning Techniques

This section presents the results of the different cleaning techniques outlined in section 3. Figures 4 and 5 show the cleaned power curves for turbine 1 only, for each of the different techniques, with and without an above rated power filter respectively. The six techniques compared were the negative power filter, the Mahalaobis Distance, the Euclidean Distance, Local Outlier Factor, Below Rated Pitching, and the Change Point Detection[7].

Figure 4 presents the cleaned power curves for the models compared with the above rated filter. The negative power looks unchanged, however it has removed the negative power values. The Mahalanobis and Euclidean distance methods look very similar, however what is most noticeable is how squared the stepping is below rated for the Euclidean distance method. This is due to the Euclidean distance itself, as it assumes a circular distribution whereas the Mahalanobis

Table 1. Percentage of Data removed by each data cleaning technique. Results bolded are from method presented in this paper.

Model	Datapoint % Difference from Original				Datapoint % Difference from Filtered			
	Turb. 1	Turb. 2	Turb. 3	Turb. 4	Turb. 1	Turb. 2	Turb. 3	Turb. 4
1) Original	0.00	0.00	0.00	0.00	N/A	N/A	N/A	N/A
2) Negative Power Filter	-17.96	-10.21	-12.19	-18.39	0.00	0.00	0.00	0.00
3) Maha	**-22.14**	**-16.10**	**-17.19**	**-22.57**	**-5.10**	**-6.55**	**-5.69**	**-5.12**
4) Maha_AR	-22.24	-16.95	-17.84	-22.83	-5.22	-7.50	-6.44	-5.44
5) Euc	-20.76	-15.47	-15.95	-21.71	-3.42	-5.86	-4.28	-4.06
6) Euc_AR	-20.86	-16.35	-16.60	-22.00	-3.54	-6.83	-5.03	-4.42
7) LOF	-18.01	-10.27	-12.22	-18.43	-0.06	-0.06	-0.03	-0.05
8) LOF_AR	-18.26	-12.45	-13.04	-19.07	-0.37	-2.49	-0.97	-0.83
9) Pitch	-22.47	-17.40	-19.47	-27.42	-5.50	-8.01	-8.29	-11.06
10) CPD & IQR	-56.14	-29.56	-34.39	-27.29	-46.54	-21.55	-25.28	-10.91

distance takes into account the shape of the distribution. Therefore, the Mahalanobis distance takes into account the angle of the distribution below rated, which is why this step effect is less noticeable. Above rated they both appear to behave the same. This step effect is caused by the Euclidean distance assuming each bin cluster is distributed as a circle, and the data is bounded on both sides by the bin boundaries.

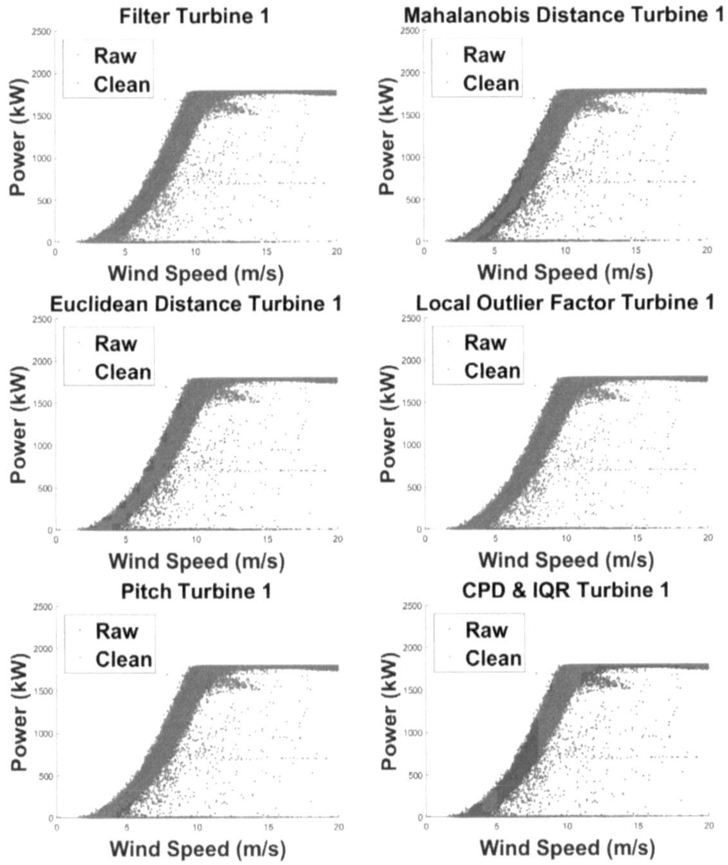

Figure 4. Each graph compares the clean and raw dataset for each cleaning technique without the above rated power filter.

The local outlier factor appears to make no difference to the data, and it is easy to mistake it for the filtered data. The below rated pitch method only seems to remove any data under the curve in the below rated regime, except for zero power. Above rated all of the outliers are included in the cleaned data, so this is not so suitable. Finally, the CPD & IQR method seems to perform very poorly. Much of the "normal" data is removed, along with other outliers. It is unknown what caused this as the method was reproduced as described in [7].

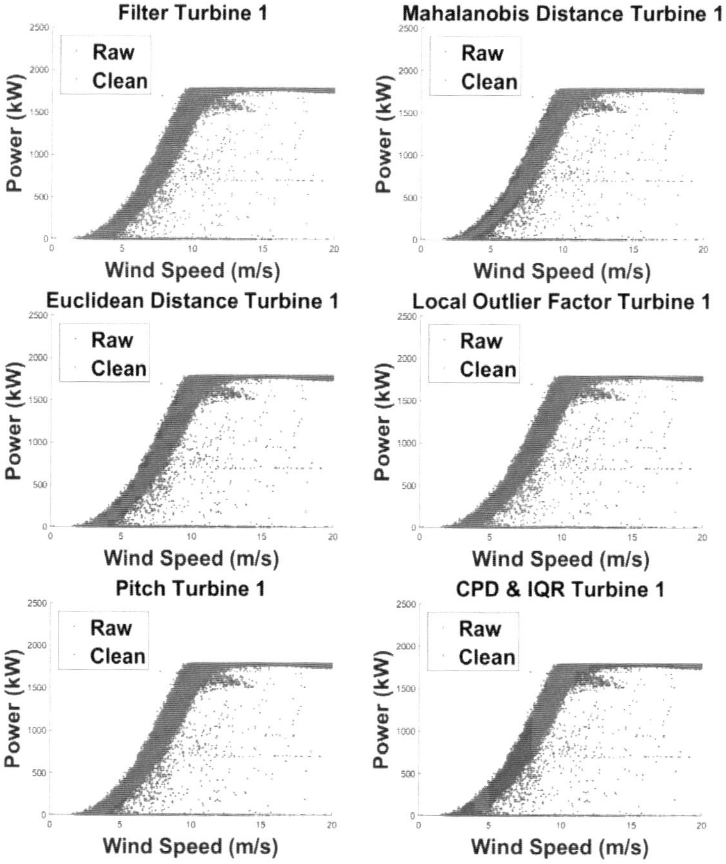

Figure 5. Each graph compares the clean and raw dataset for each cleaning technique with the above rated power filter.

Figure 5 presents the cleaned power curves using the above rated filter for all models other than the negative power filter method. For the Mahalanobis and Euclidean distance methods, the effect is much the same as discussed in the previous section. Any outliers beloww 1600 kW are removed, including any curtailment behaviour there might be. Local Outlier Factor and Pitch have above rated outliers removed, and improve their cleaning abilities over not using the above rated filter. There appears to be no discernible difference for the CPD & IQR method.

Overall, when the above rated filter is applied, and when it is not, the Mahalanobis method appears to retain more of the "normal" data than other techniques while removing as much abnormal data as possible. The technique itself is rather simple and user friendly, as it does not require anything beyond data binning and distance calculation. This is similar to the Euclidean distance, however the below rated curve is changed too much. Local outlier factor and the below rated pitch method are similarly user friendly, however their performance is unsatisfactory.

The two most successful techniques are the Euclidean and Mahalanobis distance methods, these both retain much of the normal data and remove all sparse abnormal data. Due to the introduction of the stepping effect below rated by the Euclidean distance, it is preferable to use the Mahalanobis distance. This technique also appears to be quite robust as the model performs similarly across all turbines examined. Further examination with models of a different turbine type would be useful to investigate whether this technique is still appropriate, and further comparison with the CPD & IQR method may be required to assess how robust the model is.

5. Conclusion

This paper presented a novel data cleaning technique for pre-processing SCADA data in wind turbine normal behaviour modelling for condition monitoring. The method presented first used a filter to remove negative power values, then a combination of binning and the Mahalanobis distance to remove outliers from the power curve. A filter for the above rated regime of the turbine was also considered.

This technique was compared for 4 turbines, and against a number of other cleaning techniques. It was found that the technique removed only a small percentage of the data, between 5 and 6%, and retained much of the normal data in the power curves. The Mahalanobis technique, combined with the above rated filter, has been shown to be the most effective of the techniques presented. This paper shows that the Mahalanobis technique is not only user-friendly and conceptually simple, it also proved to be the best model examined. The model was also shown to be quite robust across all turbines, however some data was removed from the normal data below rated and potentially this could be solved by changing the threshold below rated.

Future work should be aimed at utilising this cleaning technique within the framework of a condition monitoring tool. This technique, combined with appropriate feature and model selection, could potentially improve condition based maintenance method.

Acknowledgements

This research was funded by EPSRC grant number EP/L016680/1.

[1] Wind Europe 2019 Wind Energy in Europe in 2019 - Trends and statistics URL `https://windeurope.org/data-and-analysis/product/wind-energy-in-europe-in-2019-trends-and-statistics/`
[2] 2019 Unplanned wind turbine repairs to cost industry $8 billion+ in 2019 URL `https://www.woodmac.com/press-releases/unplanned-wind-turbine-repairs-to-cost-industry-8` $- billion - in - 2019/$
[3] Carroll J, McDonald A and McMillan D 2016 *WIND ENERGY* **19** 1107–1119 ISSN 11283602
[4] Faulstich S, Hahn B and Tavner P 2011 *Wind Energy* **14** 327–337
[5] Pinar Pérez J M, García Márquez F P, Tobias A and Papaelias M 2013 *Renewable and Sustainable Energy Reviews* **23** 463–472 ISSN 13640321 URL `http://dx.doi.org/10.1016/j.rser.2013.03.018`
[6] Nielsen J J and Sørensen J D 2011 *Reliability Engineering and System Safety* **96** 218–229 ISSN 09518320 URL `http://dx.doi.org/10.1016/j.ress.2010.07.007`
[7] Shen X, Fu X and Zhou C 2019 *IEEE Transactions on Sustainable Energy* **10** 46–54 ISSN 19493029
[8] Yuan L, Qiujuan H, Yi Y and Zuoxia X 2018 Abnormal State Analysis of Wind Turbines Based on the Power Curve *2018 International Conference on Power System Technology (POWERCON)* (Guangzhou: IEEE) pp 4135–4142 ISBN 9781538664612
[9] Su Y, Chen F, Liang G, Wu X and Gan Y 2019 *IEEE International Conference on Robotics and Biomimetics, ROBIO 2019* 1198–1203
[10] Hu Y, Xi Y, Pan C, Li G and Chen B 2020 *Renewable Energy* **146** 2095–2111 ISSN 18790682 URL `https://doi.org/10.1016/j.renene.2019.08.043`
[11] Long H, Sang L, Wu Z and Gu W 2020 *IEEE Transactions on Sustainable Energy* **11** 938–946 ISSN 19493037
[12] Wu J H, Shao Z G and Yang S H 2020 *Journal of Physics: Conference Series* **1639** ISSN 17426596
[13] Wang Z, Wang L and Huang C 2021 *IEEE Transactions on Instrumentation and Measurement* **70** ISSN 15579662
[14] Lin Z, Liu X and Collu M 2020 *Electrical Power and Energy Systems* **118** 105835 ISSN 0142-0615 URL `https://doi.org/10.1016/j.ijepes.2020.105835`
[15] Bao Y, Pan D, Wang X, Liao L and Yang Q 2017 Least-square B-spline Approximation Based Wind Turbine Power Curve Modeling *2017 Chinese Automation Congress* pp 6711–6716
[16] Yuan T, Sun Z and Ma S 2019 *Energies* **12** ISSN 19961073
[17] Xu X, Lei Y and Zhou X 2018 A LOF-based method for abnormal segment detection in machinery condition monitoring *2018 Prognostics and System Health Management Conference (PHM-Chongqing)* (IEEE) pp 125–128

Transferability of site-dependent wind turbine performance predictions using machine learning

Florian Hammer, Sarah Barber

University of Applied Sciences Rapperswil, Oberseestrasse 10, 8640 Rapperswil, Switzerland

E-mail: florian.hammer@ost.ch

Abstract. Within this work, machine learning models of site-specific machine learning models of wind turbine power curves of the Beberibe Wind Farm in Brazil, which consists of 32 turbines and one met mast, were developed. Previous work already showed that machine learning models taking into account site-specific effects can increase power prediction accuracy of single wind turbines by a factor of three compared to the standard power curve binning method. The main goal was to investigate the transferability of these models through power output predictions of various turbines depending on the distance from the met mast. It was found that transferring models within a wind farm is possible, but a decrease of prediction accuracy by up to 30% in certain cases could be observed. Neither the combination of various turbine data, nor the incorporation of site-specific data had an apparent effect on the transferability performance. It was thus concluded that a further investigation is needed, where a larger and more distributed subset of turbines should be used.

1. Introduction

Site-specific conditions such as turbulence intensity and shear can have a significant effect on the power output of turbines. It has been shown in previous studies that these effects can change the power output by more than 10% for a given wind speed [1, 2, 3, 4, 5, 6]. When planning a wind turbine (WTG), or even a whole wind farm project, these differences can have a tremendous influence on the estimation of the Annual Energy Production (AEP) and thus on the Levelized Cost Of Electricity (LCOE) and the Return on Investment (ROI). Clifton et al. [7] estimated a lost income of $1,000 per year per turbine based on a 1.5 MW wind turbine and an underestimated rated power of just 1%.

As the commonly used 'method of binning', standardised in the IEC-61400-12 [8], does not consider site-specific conditions apart from the wind speed, it can lead to inaccuracies and therefore higher financial risks. In recent years, power curves have been improved by incorporating machine learning algorithms that use site-specific input data such as turbulence intensity and shear [7, 9] and it has been shown that power curve prediction accuracy can be increased by a factor of three.

The values of these site-specific conditions are commonly obtained through met mast measurements within a wind farm. However, usually only a limited number of met masts at specific positions within a wind farm are set up, which can span several kilometers in length. It is therefore of interest of how well machine learning models, trained with this kind of data, can be transferred to different locations within a wind farm that are not close to met masts.

Therefore, in this work, power curve machine learning models based on measured wind speed and power data of wind turbines within a wind farm together with turbulence intensity and shear values obtained from a nearby met mast are developed. Their transferability within the wind farm is then tested through power output predictions of turbines located at various positions within the wind farm.

2. Measurement Data

The measurement data used in this work was based on SCADA data of 13 Enercon E-48 wind turbines with a hub height of $75\,m$ and a IEC-compliant met mast with a height of $100\,m$. Each measured point represents a ten-minute averaged value, which were recorded between August 2013 and July 2014. The wind speed and the power output data were obtained at the nacelle of each turbine and the turbulence intensity (TI) and the shear factor (α) were determined based on the met mast data. As the available wind turbine data was restricted to the ten-minute averaged wind speed and power output, the TI at the wind turbine locations could not be established and was obtained instead at the met mast. For the calculation of TI the wind speed and the standard deviation of the wind speed at met mast height $80\,m$ were used. The shear factor α was determined based on the power law assumption at met mast heights $40\,m$ and $100\,m$.

The turbines and the met mast are part of the Beberibe Wind Farm ('UEBB') in Brazil [10], which consist of 32 WTGs in total. Figure 1 shows a map of the site, with the three turbines used for the model development marked by red circles, the met mast in blue and the ten turbines for the transferability assessment in green. The prevailing wind on the site blows in the east-west direction.

Figure 1. The Enercon E-48 wind turbines at the Beberibe Wind Farm ('UEBB') in Brazil.

Figure 2 shows the environmental conditions at the met mast in terms of average wind speed, turbulence intensity, TI, and the shear factor, α. On the diagonal part of the matrix the histograms are depicted, whereas the lower matrix shows scatter plots. As can be seen, for lower wind speeds TI and α are much more broadly distributed, reaching high turbulence values and strong wind shear both positive and negative. Towards higher wind speeds the turbulence intensity decreases and mainly slightly positive shear factors are prevalent.

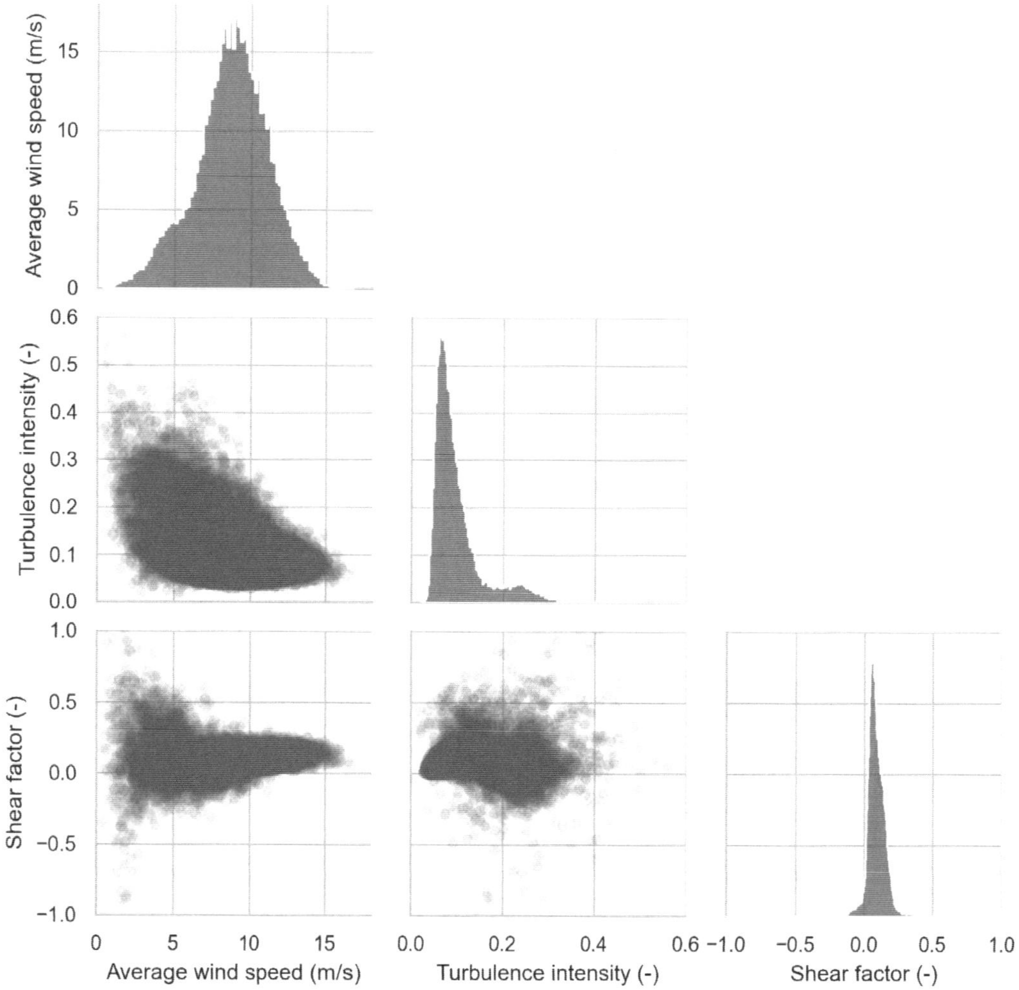

Figure 2. Histograms (diagonal) and scatter plots (lower matrix) showing the distributions and relationships, respectively, of the average wind speed, the turbulence intensity, TI, and the shear factor, α.

Turbines *WTG 12*, *WTG 13* and *WTG 14* were chosen for the power curve models due to the proximity to the met mast. The evaluation turbines were chosen for various distances from the met mast, ranging from close proximity to the met mast to distances of 2.6 km apart, which helps to better paint a picture of the transferability. A distance matrix is presented in Figure 3, showing the Haversine distance - the shortest distance on a sphere - between two points.

The power curves of turbines *WTG 12*, *WTG 13* and *WTG 14* are shown in Figure 4 and the total number of available samples per turbine are listed in Table 1.

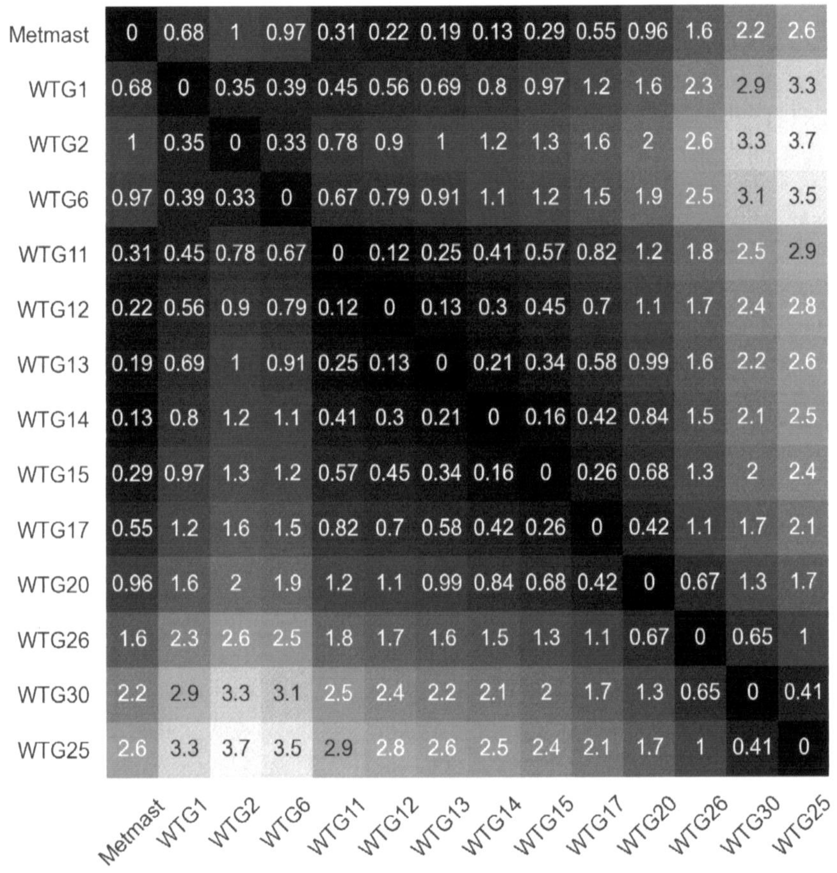

Figure 3. Distance matrix of selected wind turbines in the Beberibe Wind Farm ('UEBB') in Brazil. The lighter the color, the further the distance (km) between two points.

Table 1. Number of samples for the used wind turbines.

	WTG 12	*WTG 13*	*WTG 14*	*Combined*
No. samples	37,202	36,678	36,664	110,544

3. Wind turbine performance predictions using the gradient boosting method

The power curve models developed within this work are based on the gradient boosting technique. More specifically, the 'extreme' version of this algorithm — implemented in the Python library *XGBoost* — was chosen, which has been successfully used for a variety of complex data problems [11]. This technique offers a computationally efficient way to easily and readily generate various power curve models.

3.1. Pre-processing

The data of the different wind turbine generators used within this work was pre-processed before using it for the model training. Firstly, power values below $1.0\,kW$ for wind speeds above the cut-in wind speed of $3\,m/s$ were filtered out. After that the data was binned into $1\,m/s$ bins

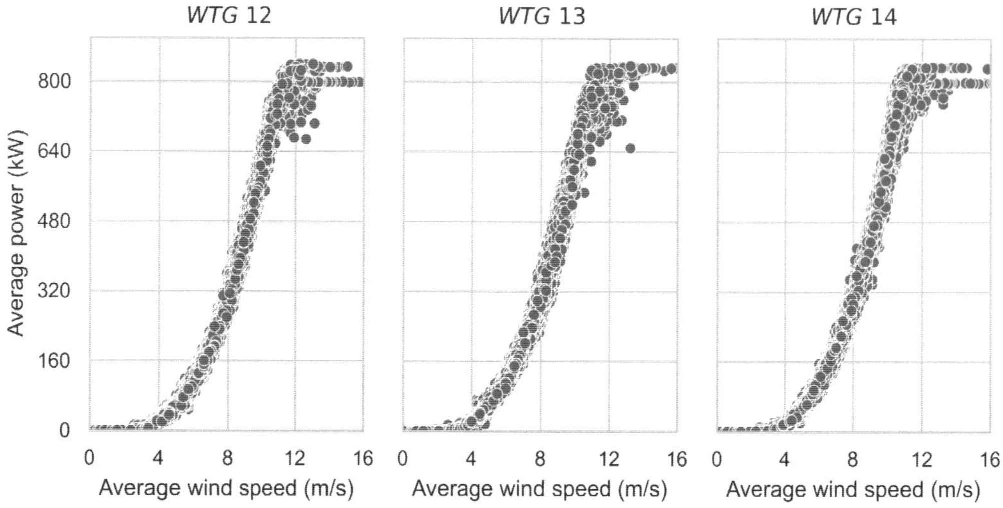

Figure 4. Measured power curves for the three wind turbines used for model development.

and the power distribution for each bin was calculated. Power values exceeding more than three standard deviations were then filtered out, resulting in a representative power curve without any outliers.

Furthermore, site-specific turbulence intensities, TI, and wind shear factors, α, taken from the met mast, were combined with the wind speed and power data of the turbines. In previous studies [7, 9] the incorporation of these site-specific conditions resulted in power curve prediction accuracy improvements by a factor of three, which is why these parameters were considered within this work as well.

3.2. Model training

The data was split into a training and a test set, based on 80% and 20% of the whole data set, respectively, where the test set is used for assessing generalization performance of the predictor. For the training, the data was further split into five equal parts, which were used for cross validation. In order to assess the performance of each model, the Root Mean Square Error

$$RMSE = \sqrt{\sum_{i=1}^{n} \frac{(\hat{P}_i - P_i)^2}{n}} \tag{1}$$

and the Mean Absolute Error

$$MAE = \sum_{i=1}^{n} \frac{\mid \hat{P}_i - P_i \mid}{n} \tag{2}$$

were calculated, where n is the number of observations, $\hat{P}_i$ is the predicted power and P_i is the actual power.

Six different power curve models were trained; for each of the three wind turbine generators *WTG 12*, *WTG 13* and *WTG 14* a model with only the wind speed as an input parameter as well as with the three input parameters wind speed, TI and α. Additionally, two models were trained based on a combined data of all three turbines, which will be denoted as the *Combined* data set in the following sections.

In order to determine the most optimal model parameters, a hyper-parameter search was conducted with the *Combined* data set, where the mean square error loss function was minimised. The parameters considered for the tuning were:

- The number of gradient boosted trees.
- The maximum tree depth.
- The minimum child weight, which controls the node splitting depending on the node's sample size.
- The learning rate.
- The booster method – in *XGBoost* the three boosters *gbtree*, *dart* and *gblinear* are available.
- The cross-validation score, which is to be optimised.

The best cross-validation score was achieved with the parameter values shown in Table 2. As

Table 2. Values of the optimised *XGBoost* model parameters.

No. of boosted trees	Max. depth	Min. child weight	Learning rate	Score
500	4	10	0.05	0.99351

for the booster, the *gbtree* booster performed better than the *dart* and the *gblinear* boosters.

Based on these optimal model parameters, learning curves for the *Combined* data set as well as the individual data sets for the turbines *WTG 12*, *WTG 13*, *WTG 14* were created, which are a means to assess whether additional data might potentially improve the prediction accuracy or not. Figures 5 and 6 depict the learning curves for one and three input parameters, respectively, where the dependence of the RMSE on the data sample size can be observed. The RMSE is determined for the training set (line) and the cross-validation set (dashed line).

As can be seen, for all wind turbine generator data sets and the combined data set, but the *WTG 14* set, the training and cross-validation losses approach each other, indicating that additional data will not further benefit the model performance. The *WTG 14* case shows a gap between the two different losses, meaning that this model does not perform as well on unseen data.

As for the case with the three inputs, wind speed, turbulence intensity TI and shear factor α, the training and cross-validation losses did not converge for all four turbine data sets. This means that additional data, if available, would potentially improve the model performance and the models in the current state suffer from overfitting and are not able to sufficiently make a generalised prediction. Even for the *Combined* data set, additional data, by incorporating more turbines in the data set, might potentially further reduce the prediction errors. This set, however, also results in the highest cross-validation and training losses.

Figures 7 and 8 show the prediction errors based on the RMSE and the MAE, respectively, for one and three input parameters. The error metrics were calculated with the test set mentioned above, which was not used for the model training itself. For all models an immediate improvement can be observed by adding the TI and α parameters, confirming the results based on simulation data mentioned above [9]. Overall, the model based on the *WTG 12* data resulted in the lowest RMSE and MAE values, whereas the model with the *WTG 14* data resulted in the highest errors. The *Combined* model performed slightly better than the *WTG 13* model. Interestingly, the *Combined* model compensated for the poor performance of the *WTG 14* model.

As the error trends are the same for RMSE and MAE, for the sake of brevity, only the MAE will be looked at in the following sections.

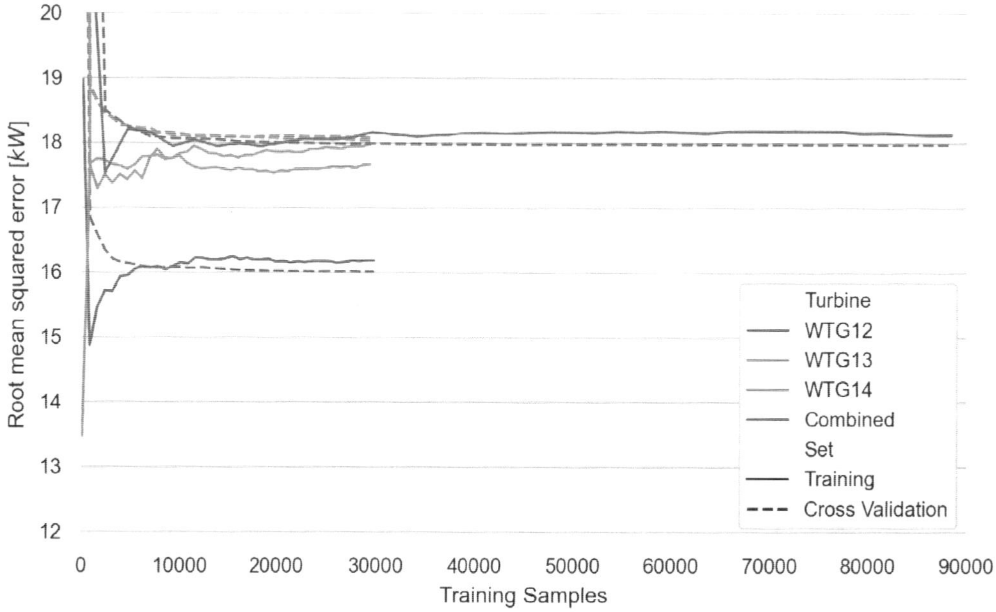

Figure 5. RMSE learning curve for the trained models based on *WTG 12*, *WTG 13*, *WTG 14* and their combination for one input.

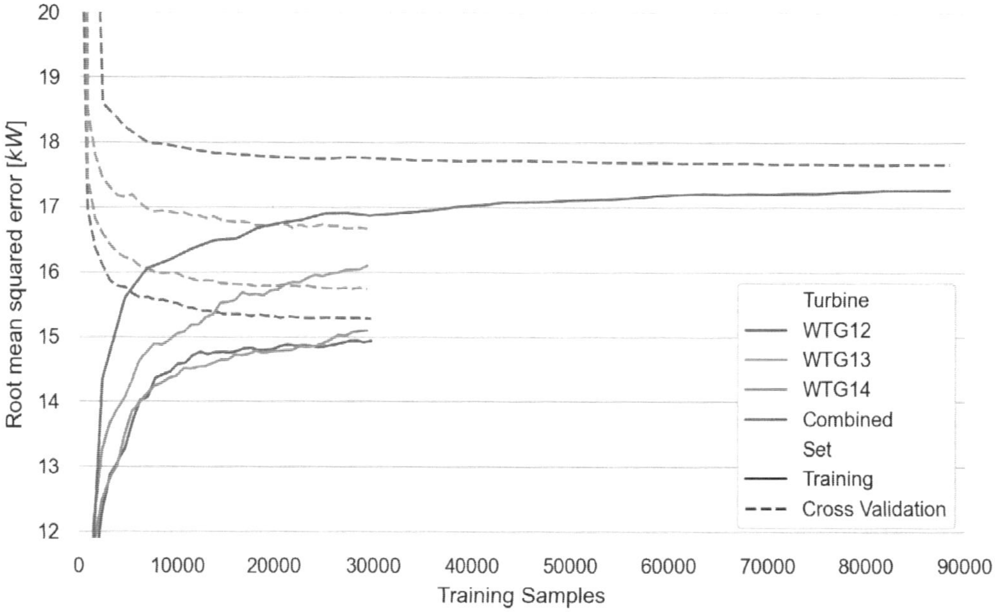

Figure 6. RMSE learning curve for the trained models based on *WTG 12*, *WTG 13*, *WTG 14* and their combination for three inputs.

4. Results

In order to assess the transferability of the eight models, data from WTGs at various distances from the met mast (Figure 1) was used to determine the prediction performance. As a result,

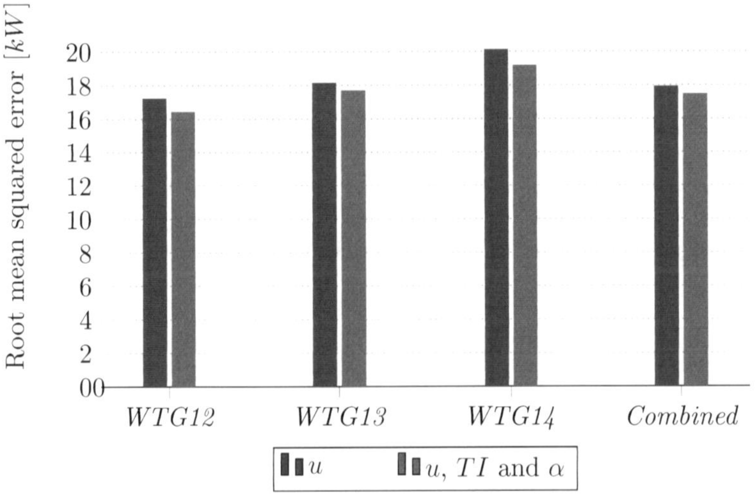

Figure 7. RMSE values for the trained models based on *WTG 12*, *WTG 13*, *WTG 14* and their combination, for one and three inputs based on their test sets.

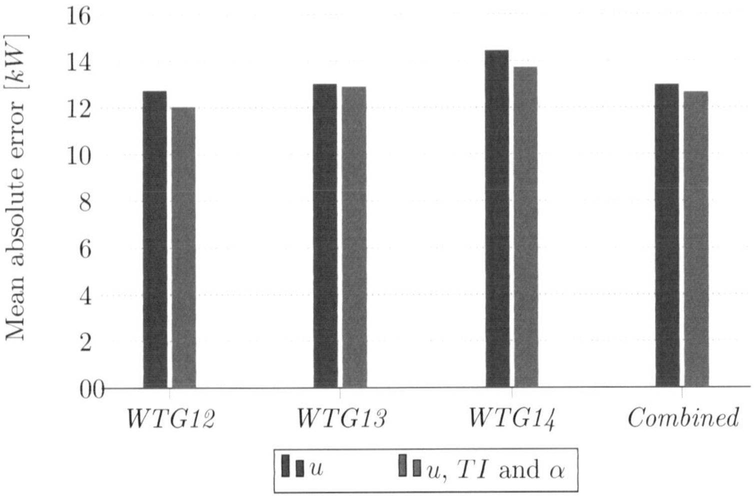

Figure 8. MAE values for the trained models based on *WTG 12*, *WTG 13*, *WTG 14* and their combination, for one and three inputs based on their test sets.

the prediction performance - the MAE - is obtained as a relationship to the distance to the met mast. The specific distances between the chosen WTGs and the met mast are shown in the distance matrix in Figure 3.

Figure 9 shows the MAE versus the distance, denoted by the respective wind turbines, for the one input parameter case, with ascending distance from left to right. As can be seen, there is no apparent trend and the prediction based on the furthest wind turbine data set, *WTG25*, which is about 2.6 km away from the met mast, is even better than some of the closer turbines. Furthermore, there does not seem to be a 'best' model between the *WTG 12*, *WTG 13*, *WTG 14* and *Combined* models for distances up to 1 km (*WTG2*) from the met mast, as each of them perform better or worse for different turbines. However, for the three furthest turbines the model based on *WTG14* performs best. This is interesting, considering that the *WTG14* model performance was worst based on its own test set, as observed in Section 3.2.

The same conclusions can be drawn for the three input parameter case, shown in Figure 10. The addition of the turbulence intensity, TI, as well as the shear factor, α, has almost no positive effect on the prediction losses. This might be due to the fact that the site-specific conditions are less relevant for the other WTGs as for *WTG 12*, *WTG 13* and *WTG 14* because of the larger distance to the met mast.

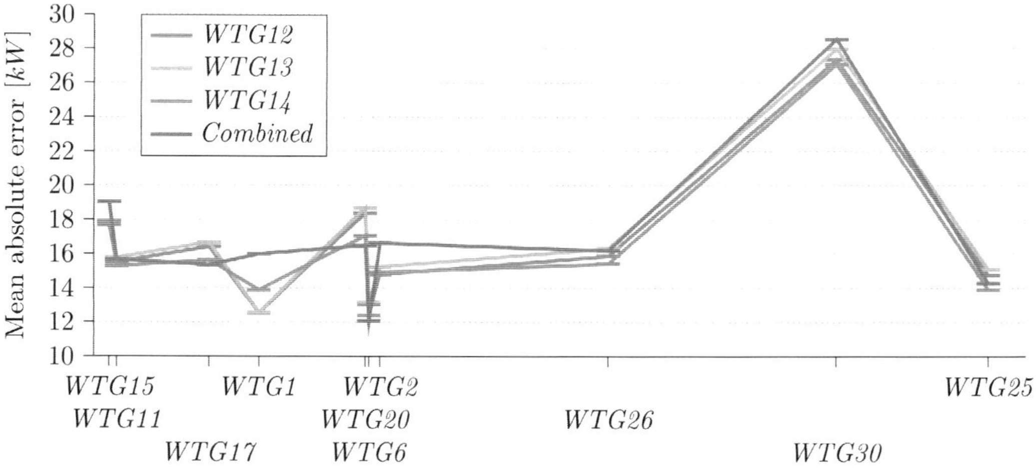

Figure 9. Transferability of the power curve models based on *WTG 12*, *WTG 13*, *WTG 14* and their combination for the one input parameter case.

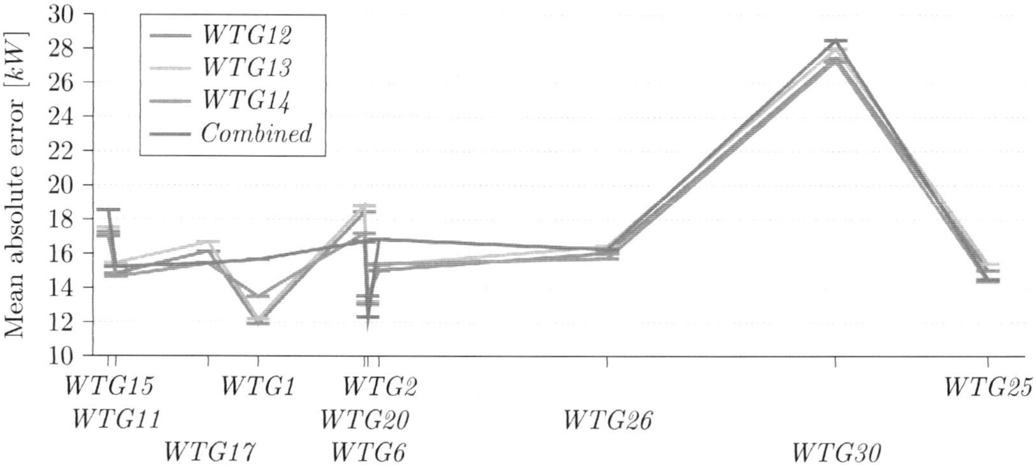

Figure 10. Transferability of the power curve models based on *WTG 12*, *WTG 13*, *WTG 14* and their combination for the three input parameters case.

Additionally, for each individual turbine shown in the previous plots (*WTG 1*, *WTG 2*, *WTG 30*, etc.) two models, with one and three input parameters respectively, based on their own data were trained and tested. The data sets were also pre-processed according to the procedure given in Section 3.1. For the training the same model parameters as listed in Table 2 were used. The model performances were then compared to the *WTG 12*, *WTG 13*, *WTG 14* and *Combined* model performances. In order to do so the differences in the prediction losses were calculated, which are shown in Figures 11 and 12. The first thing to notice is that each individual turbine model performs better than the *WTG 12*, *WTG 13*, *WTG 14* and *Combined* models used for predictions on these turbines. However, for some turbines the differences are very small, showing good model performance on unseen data. Especially the *WTG 14* model stands out when considering the three furthest turbines *WTG 26*, *WTG 30* and *WTG 25*, indicating great transferability for the given data. As already observed above, the *Combined* model shows mixed prediction performance for all considered turbines.

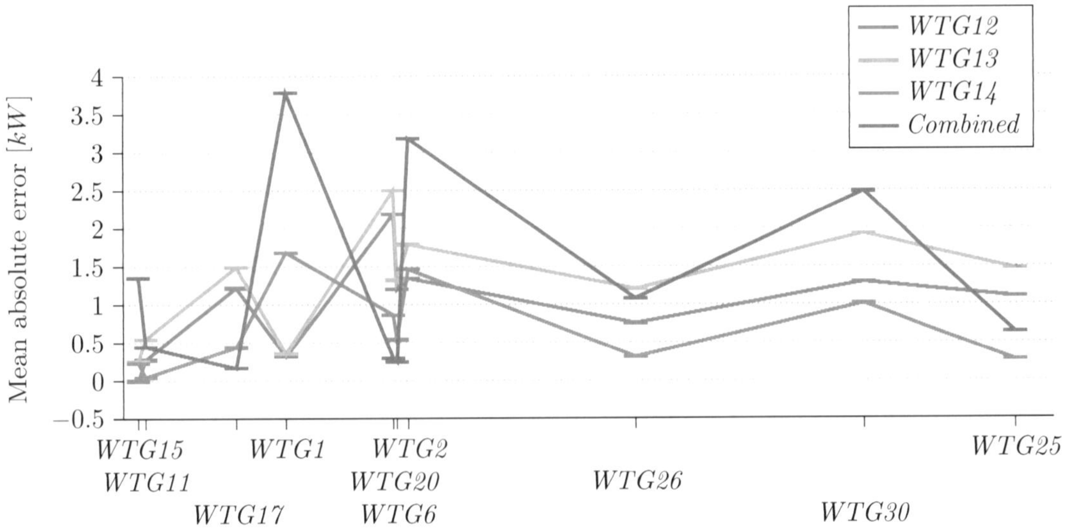

Figure 11. Differences between the prediction losses based on the *WTG 12*, *WTG 13*, *WTG 14* and *Combined* models and the prediction losses for the individual turbine models for the one input parameter case.

5. Conclusions

Based on these results, some important conclusions can be drawn. Firstly, incorporating site-specific conditions such as turbulence intensity, TI, and the shear factor, α, has a positive effect on the prediction performance on single turbines, reducing the RMSE by almost 5% when considering the boosted tree models. However, this was only the case for the data based on the trained models and not for unseen data of wind turbines some distance apart. Secondly, more training data is needed when site-specific conditions are to be considered, as revealed by the learning curves. The bundling of data of various wind turbines and combining them in one model helped tackle this problem and even compensated for poor model performance, as seen for *WTG14*. Hence, it is advisable to combine data sets of several turbines in order to achieve a more robust result. However, these advantages were not observed in terms of transferability. Thirdly, transferring models within a wind farm is possible, but a decrease of prediction accuracy by up to 30% in certain cases could be observed. Furthermore, neither the combination of turbine data, as in the case of the *Combined* model, nor the incorporation of site-specific data, had an apparent effect on the transferability performance and further analysis is needed.

This leads to the overall conclusion that the transferability performance needs to be further investigated with a larger and more distributed subset of turbines, which is part of our ongoing

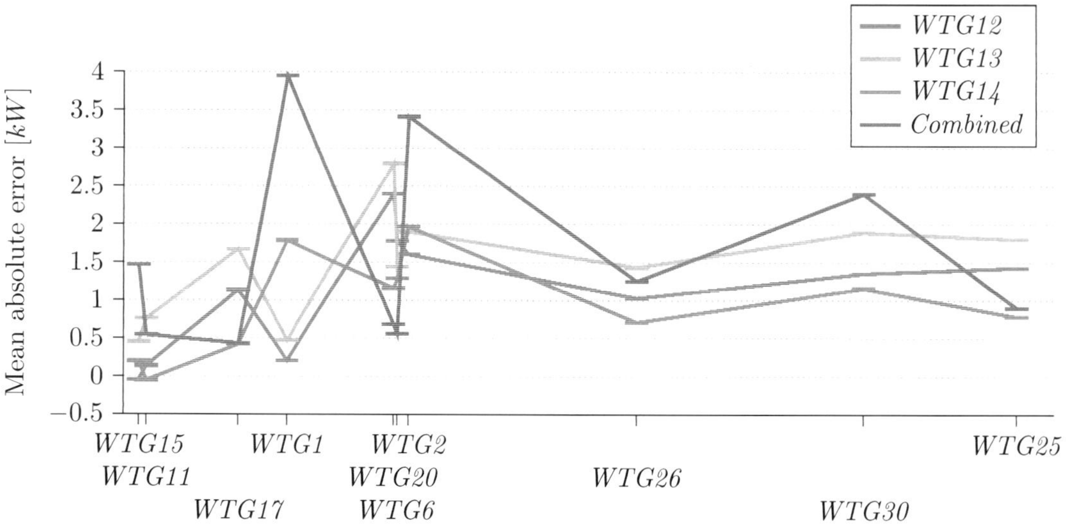

Figure 12. Differences between the prediction losses based on the *WTG 12*, *WTG 13*, *WTG 14* and *Combined* models and the prediction losses for the individual turbine models for the three input parameters case.

work. Other site-specific variables like air density and wind direction, if available, should be considered as well in order to improve model accuracy [1]. At certain wind directions, for example, turbine wakes will affect neighboring turbines and hence have a direct impact on the power production. Wake effects might also be one possible explanation for the observed prediction performance behavior in relation to the distance. Another important variable influencing the power production of a turbine is the yaw error, occurring when the wind direction and the rotor axis are not perpendicular to each other [12]. Transferability considerations could also benefit from further wind farm analysis techniques and methods, like spatio-temporal analysis, in order to gain more insights into the flow patterns and relationships between turbines within a wind farm.

References

[1] Lee G, Ding Y, Genton M G and Xie L 2015 *Journal of the American Statistical Association* **110** 56–67 ISSN 0162-1459, 1537-274X URL `http://www.tandfonline.com/doi/full/10.1080/01621459.2014.977385`
[2] Antoniou I, Pedersen S M and Enevoldsen P B 2009 *Wind Engineering* 449–468
[3] Hedevang E 2014 *Wind Energy* 173–195
[4] Vanderwende B J and Lundquist J K 2012 *Environmental Research Letters* 034035 ISSN 1748-9326
[5] Wagner R, Courtney M, Gottschall J and Lindelow-Marsden P 2011 *Wind Energy* 993–1004
[6] Wharton S and Lundquist J K 2012 *Environmental Research Letters* 014005 ISSN 1748-9326
[7] Clifton A, Kilcher L, Lundquist J K and Fleming P 2013 *Environmental Research Letters* **8** 024009 ISSN 1748-9326
[8] IEC International Electrotechnical Commission 03032017 Wind energy generation systems - part 12-1: Power performance measurements of electricity producing wind turbines
[9] Barber S and Nordborg H 2020 *Journal of Physics: Conference Series* **1618** 062003 URL `https://doi.org/10.1088/1742-6596/1618/6/062003`
[10] Sakagami Y, Pedro A, Santos A, Haas R, Passos J and Taves F 2015 *EWEA Annual Event 2015*
[11] Chen T and Guestrin C 2016 *ArXiv e-prints*
[12] Wan S, Cheng L and Sheng X 2015 *Energies* **8** 6286–6301 ISSN 1996-1073 URL `https://www.mdpi.com/1996-1073/8/7/6286`

Artificial intelligence-based condition monitoring and predictive maintenance framework for wind turbines

Janine Maron[1*], Dimitrios Anagnostos[1], Bernhard Brodbeck[1], Angela Meyer[2]

[1] WinJi AG, Badenerstrasse 808, 8048 Zurich, Switzerland [1*]
[2] Bern University of Applied Sciences, Quellgasse 21, 2501 Biel, Switzerland
*Corresponding author: janine.maron@win-ji.com

Abstract. The global wind power capacity continues to grow at a fast pace. However, the profit margins from wind power are being compressed in many countries. Thus, many wind farm owners seek to reduce their operational expenses, including those for maintenance work. In this study, an artificial intelligence-based condition monitoring and predictive maintenance framework for wind turbines is presented. The purpose of this framework is the automated early detection of operational faults in wind turbine systems and subsystems. The early detection of anomalies enables further diagnosis, condition-based maintenance and better planning of repairs. It can prevent consequential damage, lead to fewer turbine downtimes and extend the service lives of the monitored turbines. We present validation results from two onshore wind farms and demonstrate 97% accuracy for a 2-month detection horizon of developing fault events that require attention from maintenance staff.

1. Introduction

While the global wind power capacity is growing at astonishing rates, the guaranteed profit margins for owners and operators are constantly being compressed due to competitive energy markets [1]. Owners and managers of such assets are exploring additional revenue creation through the optimisation of their operation and maintenance activities. Especially for remote wind farms, such as offshore sites, predictive maintenance can greatly affect CAPEX spend, especially when performed frequently and remotely, avoiding expensive on-site visits [2-4].

The aim of this work is to propose and evaluate a framework for providing automated condition-based maintenance decision support for wind turbines and their electromechanical subsystems based on artificial intelligence algorithms. The purpose of this framework is to detect developing faults in wind turbine systems and subsystems as early and accurately as possible. The early detection of anomalies enables condition-based maintenance and better planning of repairs. This prevents consequential damage, leads to fewer turbine downtimes and extends the service life of a turbine [5-7].

This paper is organized as follows. Section 2 introduces the datasets used in this study and section 3 presents the methods developed. Our results are discussed in section 4. Finally, section 5 outlines our conclusions.

2. Datasets

In this work the condition monitoring data from the SCADA system of two commercial European onshore wind farms have been analyzed. These are variable-speed three-bladed horizontal axis turbines with pitch regulation. Table 1 provides an overview of the turbines' technical specifications.
The data have been anonymized for this study. Ten-minute average values of condition variables from the supervisory control and data acquisition (SCADA) system served to train and test the model as specified below. The data represent approximately 30 months of SCADA data from a wind farm with four turbines (wind farm 1) and 4.5 years of SCADA data from a second wind farm with eight turbines (wind farm 2). The data comprise the rotor and generator speeds, active power production, currents and the temperatures of multiple drivetrain components, including gear bearings, generator windings and hydraulic oil temperatures, amongst other condition variables.
Additionally, independent information on repairs and technical problems of individual turbines in wind farms 1 and 2 has been evaluated to compare with the results of the proposed method.

In this work the results for one wind turbine of each farm are being presented. In wind farm 1 all turbines show similar results as the one presented in this work. In wind farm 2, serious anomalies were only recognized for the turbine presented here, the analyzed technical components of all other turbines appear to be healthy.

Table 1. Turbine specifications

	Wind farm 1	Wind farm 2
OEM	Vestas	Nordex
Installed Power	2000 kW	2300 kW
Cut-in wind speed	4 m/s	4 m/s
rated wind speed	12 m/s	13 m/s
Cut-out wind speed	25 m/s	25 m/s
Hub height	80 m	100 m
Rotor diameter	90 m	90 m

3. Methods

The core idea of the proposed method is to create a digital model that predicts the normal behaviour of the system of interest based on its operating conditions and that detects significant deviations of predicted and actual behaviour. A deviation from this model can potentially indicate anomalies in the system [3,6,7]. The method can be separated into four steps: preprocessing, modeling, anomaly detection and postprocessing.

The first part of the algorithm involves preprocessing of the above datasets. Data preprocessing deals with rigid data cleaning and transformation of real measured data. This necessitates data cleaning and transformations. Filtering is applied to arrive at clean and meaningful datasets. Conditions of filtering relate to:

- data containing blank values for one of the input or the output variables of the model
- times when a turbine is not producing, comprising the following cases
    - the turbine is not producing due to low wind
    - the turbine is not producing despite sufficient wind. In this case the turbine is in a downtime, which can be caused by e.g. maintenance, grid or environmental curtailments.
- times when the power output of the turbine is reduced from the outside (e.g. grid curtailment, noise mode)
- times within two-hour phase after a shutdown, because technical parameters are not in their normal range shortly after a shutdown
- outliers and corrupted data.

After cleaning, the data is transformed to be comparable on the same scale. All technical parameters as well as power, wind speed and rotor speed have been log-transformed to this end.

In order to train the models, a training data set was defined based on a multi-months time period in which the turbines demonstrated no unusual operating behaviour and in which no maintenance needs occurred to the best of our knowledge. When there is a software update or another drastic change in the behaviour of the turbine, the models need to be retrained on the new setup. Furthermore, all important characteristics, for example seasonality, must be covered in the data and the dataset needs to contain enough information in order to prevent overfitting. This was ensured by the choice of the training dataset. The first year of the dataset was selected as training data. In case of a continuous anomaly detection process, a periodic retraining of the model is initiated.

Multiple machine learning algorithms were compared for training models of the normal operating behaviour of the turbines, including a random forest algorithm [8], support vector machines [9] and a gradient boosting algorithm [10]. The model that provided the best performance in terms of model training time and robustness was trained with a gradient boosting algorithm. Compared to the other training methods like the random forest algorithm and support vector machines, the model trained with the boosting algorithm achieved similar accuracies but was ten times faster in computation.

A separate model was trained for each turbine and subsystem sensor. We refer to a discussion of multi-target normal behaviour models in [11]. In the present study, the measures of wind speed and environmental temperature were taken as input variables to train the models of the considered turbine subsystems. With these two input variables, a model with sufficiently accurate fault detection accuracy could be created for the majority of the available sensors in the turbine subsystems. The input variables were limited to these two predictors due to their explanatory power and their high availability in most of the wind turbine data sets.

Finally, in order to declare data points as anomalies, two threshold conditions were introduced with regard to the residuals and the occurrence frequency of large residuals. First, the residual threshold was selected as the 99 %-quantile of the residual distribution to fit the distribution and following the assumption that the data in the training set represent the normal behaviour of the turbine. We focussed on the residuals as the difference between the prediction and the actual value of the target variable. Whenever a residual exceeded this threshold, it was flagged as a potential anomaly. A fault was flagged only if anomalies occurred frequently enough within a certain period of time. A frequency-related threshold was introduced to this end. It is used to distinguish actual long-term anomalies from single isolated deviations.

For a fault to be detected, this second threshold requires that a given minimum number of potential anomalies occur within a rolling time window of 24 hours. We found that setting this second threshold to at least 40 potential anomalies within 144 data points (corresponding to a 24-hour window) provides a satisfactory trade-off between too late detection and too many false positive alarms in practice. Based on the residual and frequency thresholds, anomalies could now be accordingly identified in the test set.

4. Results

The extensive data collected by WinJi[1] through its digital services have been exploited in this work in order to develop and test an accurate fault detection scheme and demonstrate its performance in commercially operating wind turbines. We have compared and benchmarked a number of state-of-the-art algorithms for the creation of normal-operation models of the main subsystems of monitored wind turbines and for the early detection of developing operating faults. The validation was performed on two wind farms in the asset portfolio monitored by WinJi. The framework enables the detection of maintenance needs in gearbox and generator anomalies approximately two months before maintenance is actually performed, as shown in Figure 1. We could also verify that the maintenance work was effective in that the detected temperature anomalies disappeared after both monitored incidents (Figure 1). The developed anomaly detection framework was tested on the turbines of wind farms 1 and 2 in the portfolio of WinJi. The anomaly detection process described in section 3 was carried out for each turbine and monitored turbine condition variables separately.

As outlined in section 3, the fault alarm criterion is based on two different thresholds using the residuals computed as difference of the observed sensor measures and the predictions by the above normal behaviour model. An alarm was raised when the 99 % quantile of the residuals distribution was exceeded for more than 6.7 hours in the respective past 24 hours.

Table 2 details the subsystems sensor variables that were examined and the root mean square errors (RMSE) of the trained models per farm.

Table 2. Overview of subsystems sensor variables and RMSE of trained model (log-transformed test data) listed per farm. In case there was no data available or the trained normal behaviour model was not satisfactory, there is no RMSE value listed. As part of the preprocessing the variables have been log-transformed before model training.

	Wind farm 1	Wind farm 2
Converter Temp [°C]	0.039	-
Gear Bearing Temp [°C]	0.1	0.059
Gearbox Temp [°C]	0.036	-
Gear Oil Sump Temp [°C]	0.046	-
Generator Speed [rpm]	0.023	0.025
Generator Journal Temp A [°C]	-	0.086
Generator Journal Temp B [°C]	-	0.097
Generator Temp 1 [°C]	-	0.060
Generator Temp 2 [°C]	-	0.0

[1] www.win-ji.com

Hydraulic Oil Temp [°C]	0.104	-
Hydraulic Pressure [bar]	0.004	-
Power [kW]	0.104	0.139
Rotor Speed [rpm]	0.024	0.033
Spinner Temp [°C]	0.144	-
Transformer Phase Temp 1 [°C]	0.127	-
Transformer Phase Temp 2 [°C]	0.113	-
Transformer Phase Temp 3 [°C]	0.126	-

For the first wind farm, anomalies are predicted for the gearbox. Three out of four turbines were flagged in October to November 2019 and 2020, as presented in Fig. 1. Maintenance was carried out in November of both years, after which the anomalies ceased.
For the second wind farm, one out of eight turbines was diagnosed with unusually increased generator operating temperatures, as shown in Fig. 2. In this case the owner could be notified in time so a generator failure and possibly nacelle fire could be prevented. Both cases provided a time window of more than one month for early intervention and prevention.
Comparing the applied algorithms, we found that models trained based on the gradient boosting algorithm [10] performed most robustly in automated unsupervised detection of faults based on the component's normal operating behaviour. Moreover, while they achieved similar accuracy in describing the normal operating behaviour, the models trained with the gradient boosting algorithm outperformed the other benchmarked algorithms in terms of the model training times. The latter is a relevant and beneficial algorithm property for scaling our approach to a large number of monitored components and wind farms as in the WinJi portfolio. The RMSE of the compared models (Random forest, SVM, Boosting) and the times in seconds needed for training on an Apple M1 processor are shown in Table 3.

Table 3. Overview of multiple machine learning algorithms for training models of the normal operating behaviour of the turbines, the achieved RMSE of the trained model (log-transformed test data) and the model training time using the example of variable generator temperature 1.

	RMSE on test dataset for variable generator temp 1 [° C]	Model training time [sec] for variable generator temp 1
Support vector machine	0.062	1320
Random Forest	0.064	2040
Gradient boosting	0.060	130

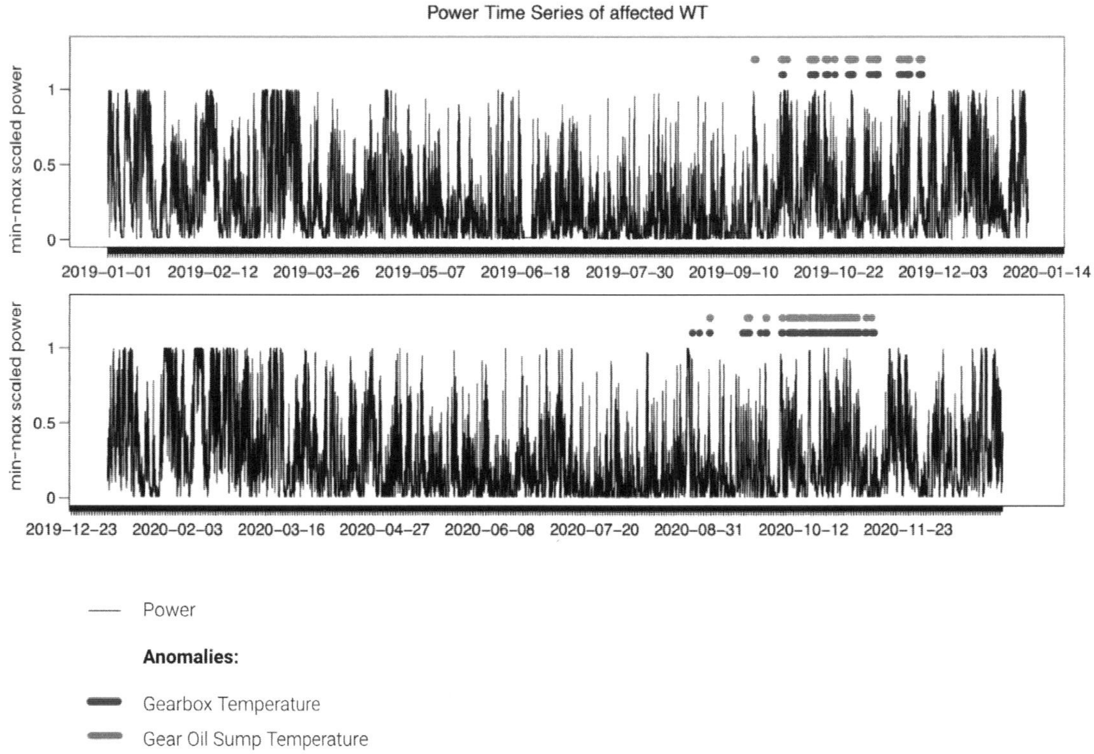

Figure 1. Active power produced by the turbine that developed a gearbox fault in wind farm 1 scaled with the min-max method to a range in [0, 1]. Anomalies detected with the developed models for a subset of 13 monitored WT condition variables, namely the gearbox temperature and gear oil sump temperature, based on 10-minute values from the SCADA system in wind farm 1.

As presented in Fig. 1, an alarm was triggered when the residuals indicated unusual operating behaviour of the respective WT component based on the defined alarm criterion. Excessive gearbox temperatures first start to evolve in Sept. 2019, indicating maintenance need in the gearbox. The alarm switches off immediately after the work has been performed on 17 Nov. 2019, indicating the effectiveness of the performed maintenance activities. A similar situation recurred in 2020 in the same gearbox, with gearbox temperature starting to rise unusually in Sept. 2020 and maintenance performed to successfully resolve the condition on 15 Nov. 2020.

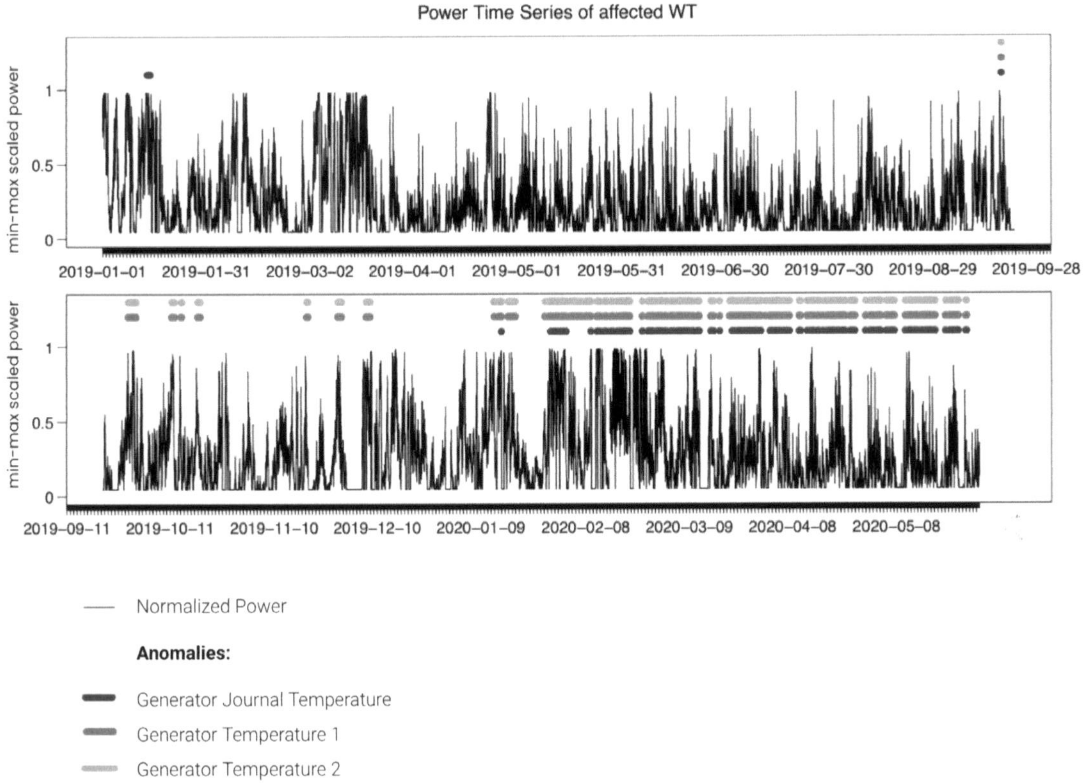

Figure 2. Demonstration of the fault detection algorithm for drivetrain subsystems of a turbine of wind farm 2.

Another result from the validation of the developed framework is the estimation of the probability that a fault causing downtime will actually appear. For this we used the following methodology:

- For each detected anomaly, we checked the error code information k days into the future for downtimes. k is the number of days following an anomaly that we examined for downtimes.
- We only considered downtimes connected to technical failure and unplanned maintenance.
- k was chosen as a rolling window with values [5, 10, 20, 30, 60]
- The Positive Predictive Value (PPV) of the framework was assessed, based on the following equation:

PPV = (Number of True Positives) / (Number of True Positives + Number of False Positives)

Fig. 3 provides an overview of the PPV values for the main systems that exhibited useful correlations for this dataset. When the proposed method detects an anomaly for the gearbox or the converter, it is almost certain (>97%) that within a month (k = 30), the wind turbine will experience a fault and stoppage. PPV for transformers is lower but still above 75% and interestingly the gear bearing temperatures do not produce a relevant result with PPV never exceeding 20%.

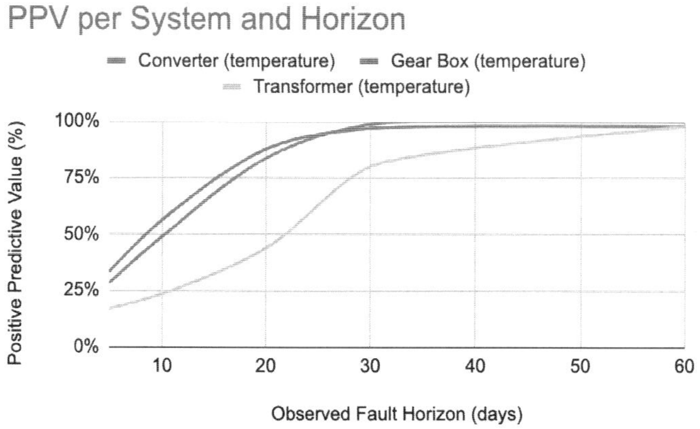

Figure 3. PPV graph vs. fault horizon. In parentheses the type of signal that was used per system.

5. Conclusions

We have presented a condition-based maintenance framework for wind turbines that can successfully detect developing faults at an early stage and assess the effectiveness of performed maintenance activities using artificial intelligence techniques. The automated early detection enables condition-based maintenance work and informed planning and scheduling of repair activities. This helps prevent consequential damage, leads to fewer turbine downtimes and can extend the service life of the monitored turbines. We have presented validation results from two commercial onshore wind farms. We also demonstrated that the framework enables verifying the effectiveness of performed maintenance work.

The presented framework is capable of manufacturer-agnostic early fault detection and allows notification of the wind farm managers and owners. This can lead to optimised operations and generation of additional value from the asset while also avoiding unplanned expenses and downtimes. Furthermore, this framework is part of a digital platform, providing transparency and insights fully remotely with minimal visits on-site. As demonstrated, such systems are highly beneficial for efficient management of large portfolios and value creation in the assets themselves.

6. Acknowledgements

This work has been partially funded by Innosuisse, project nr. 31533.1 IP-EE.

References

[1] McInnis, D., M. Capezzali, 2020, Managing Wind Turbine Generators with a Profit Maximized Approach, Sustainability 12, doi: 10.3390/su12177139.

[2] International Renewable Energy Agency, 2019, Renewable power generation costs in 2019, International Renewable Energy Agency.

[3] Meyer, A., B. Brodbeck, 2020, Data-driven Performance Fault Detection in Commercial Wind Turbines, Proceedings of the 5th European Conference of the Prognostics and Health Management Society 2020, Turin, Italy, ISBN 978-1-93-626332-5.

[4] Fischer, K., D. Coronado, 2015, Condition monitoring of wind turbines: State of the art, user experience and recommendations, VGB PowerTech 2015(07):51-56.

[5] Hameed, Z., Y. Hong, Y. Cho, S. Ahn, C. Song, 2009, Condition monitoring and fault detection of wind turbines and related algorithms: A review, Renewable and Sustainable Energy Reviews 13, doi: 10.1016/j.rser.2007.05.008.

[6] Zaher, A., S. McArthur, D. Infield, Y. Patel, 2009, Online wind turbine fault detection through automated SCADA data analysis, Wind Energy, 2009, doi: 10.1002/we.319.

[7] Tautz-Weinert, J., S. Watson, 2017, Using SCADA data for wind turbine condition monitoring – a review, IET Renew. Power Gener. 11, doi: 10.1049/iet-rpg.2016.0248.

[8] Breiman, L., 2001, Random Forests, Machine Learning 45: 5–32, doi: 10.1023/A:1010933404324.

[9] Drucker, H., C. Burges, L. Kaufman, A. Smola, V. Vapnik, 1996, Support Vector Regression Machines, Proceedings of the 9th International Conference on Neural Information Processing Systems, Denver, CO, USA, December 2-5, 1996.

[10] Chen, T., C. Guestrin, 2016, XGBoost: A Scalable Tree Boosting System, Proceedings of the 22nd ACM SIGKDD International Conference on Knowledge Discovery and Data Mining (KDD '16), Association for Computing Machinery, New York, NY, USA, 785–794, doi: 10.1145/2939672.2939785.

[11] Meyer, A., 2021, Multi-target normal behaviour models for wind farm condition monitoring, Applied Energy, Elsevier, doi: 10.1016/j.apenergy.2021.117342.

Journal of Physics: Conference Series **2151** (2022) 012008 doi:10.1088/1742-6596/2151/1/012008

Investigation of the Measurability of Selected Damage to Supporting Structures of Wind Turbines

J Rupfle[1], A Emiroglu[2], C U Grosse[1]

[1] Chair of Non-destructive Testing, Technical University of Munich, Franz-Langinger-Str. 10, 81245 Munich

[2] Chair of Structural Analysis, Technical University of Munich, Arcisstr. 21, 80333 Munich

E-mail: johannes.rupfle@tum.de, grosse@tum.de

Abstract. The present work deals with a measurability study of damage to supporting structures of wind turbines. The examined hybrid tower consists of a prestressed concrete part with an attached steel section on top. Principal focus is on developing a data preprocessing and analysis concept aiming to investigate the measurability of the two selected failure cases *missing pretension of tendons* and *fatigue based change of Young's modulus in reinforcement concrete*. A simulation deduced out of real world mass and thrust forces provides a comparison with measured data based tower displacements, natural frequencies and mode shapes. The measurability of given deviations is investigated by means of virtual sensors, derived from the measurement setup of a test facility measuring acceleration, velocity, displacement and strain on different positions of the wind turbine tower, aiming the development of detection methods of such cases of damage for condition monitoring systems. The simulation input can be given by strain-based external forces or vibration-based displacements.

1. Introduction

Condition Monitoring has established itself in the past years as an instrument for damage detection. The project *IM Wind* as well as the previous project *MISTRALWIND* are focused on long-term monitoring of wind turbines [1]. The present work is concerned with measurement techniques for the detection of damage to the supporting structures of wind turbines. Using a construction plan and statics based model, the measurability of missing pretension of tendons and fatigue based change of Young's modulus in reinforcement concrete as failure cases influencing Eigenfrequencies, damping values and mode shapes is discussed.

A comprehensive literature research and reflected discussions with experts led to a number of possible damages to supporting structures of wind turbines. An overview of all considered damages is given in figure 1. In case of the within the framework of the project equipped hybrid tower, damage is mainly concentrated on connecting elements [2] [3]. A group of experts around the IM Wind Project carried out a final selection of the damage cases to be considered in the evaluation.

As the hybrid tower consists of a prestressed concrete part, tendons are used to avoid stress in reinforcement concrete to avoid damage. The pretensioning force is adjusted with a hydraulic

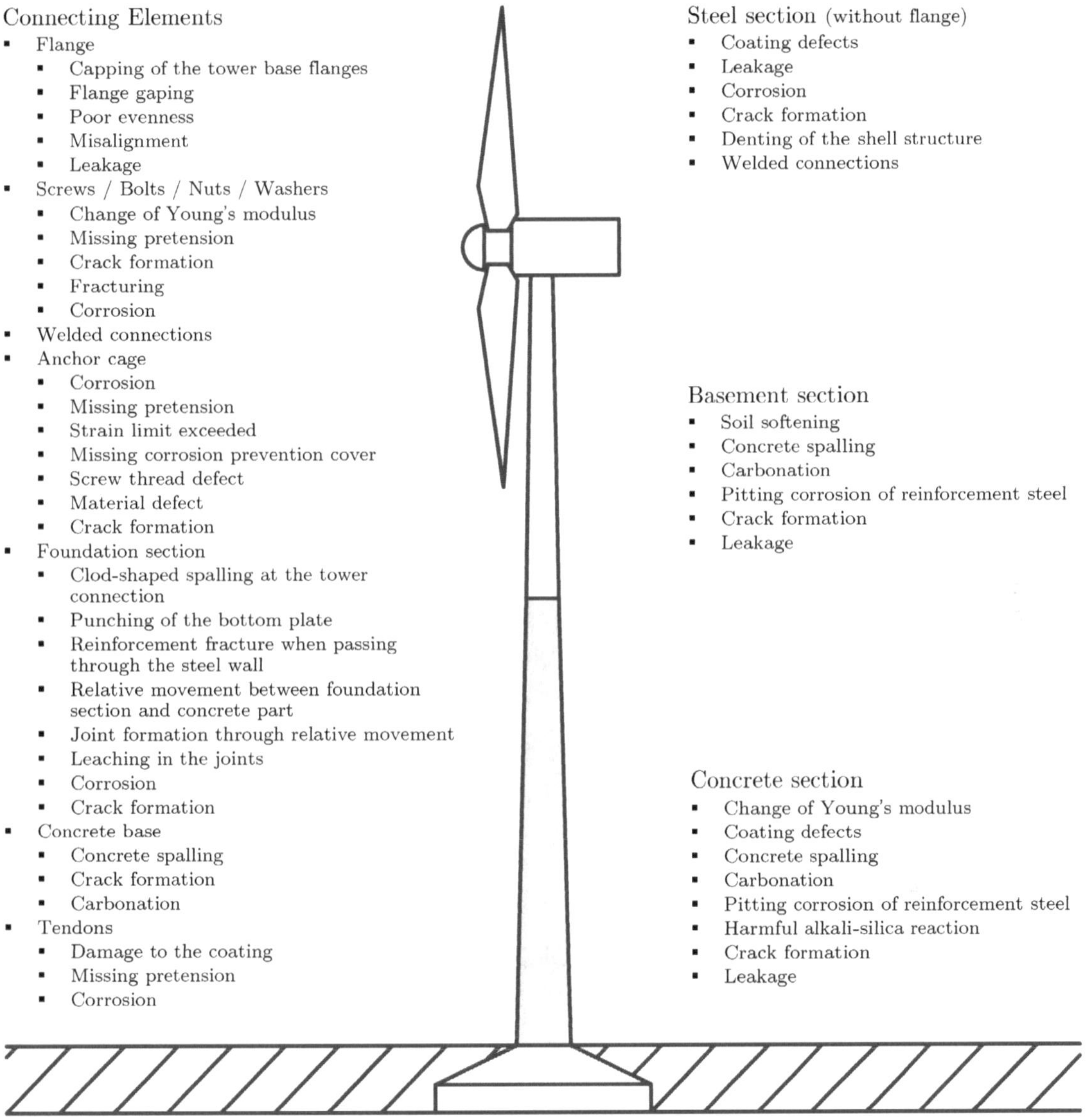

Figure 1. Possible damage considered in the evaluation process divided into subsections.

press and must be checked at regular intervals. Since a missing prestressing force will have a direct effect on the overall stiffness of the tower, it can be assumed that this can be recorded by measuring techniques.

Furthermore, degradation of reinforced concrete is investigated. As examined by Musiał [4] and Owen [5], the Young's modulus of reinforced concrete can be determined by measuring the Eigenfrequencies. Vibrational data can be used to monitor small changes in stiffness of reinforced concrete associated with slow degradation under cyclic load.

To gain knowledge about the measurability of damage in form of a reduced prestressing force, acceleration measurements of an equipped wind turbine are compared to simulation data based on mass and thrust forces derived from strain measurements combined with Supervisory Control and Data Acquisition (SCADA) data.

An essential element of the presented data preprocessing procedure is the Untwisting Cable Event (UCE) of a wind turbine. In this event, the twisted cable bundle in the tower of the turbine is untwisted to prevent damage. This specific condition is used to calibrate and synchronize measurement equipment.

2. Measurement Setup

The test facility is equipped with a variety of different measurement instruments to allow testing and validation of various approaches to a condition monitoring system and to investigate their accuracy. The sensors are distributed over five levels from the foundation up to the top of nacelle. All sensor axes are converted to a tower-fixed cylindrical coordinate system x-y-z by means of composed elemental rotations with an intrinsic sequence X-Y'-Z''.

The main concept of the condition monitoring system is based on acceleration measurements, distributed over a height from 0 m to 139 m. They are complemented by strain gauges at heights of 81 m and 139 m, which are paired with temperature sensors. In addition to the strain gauges, a fiber optic measurement system with a total of eleven distributed strain and temperature sensors is located in the tower base area. A further seismometer is also located in the tower base. Tilt sensors are installed in the tower base and at a height of 139 m. In addition, a Real Time Kinetic (RTK) module is located on top of the nacelle, which provides a maximum resolution in the centimeter range and data with a sampling rate of 14 Hz. As the MISTRALWIND project shows, the Eigenfrequencies of the tower as well as from the rotor blades and influences of the rotor speed can be determined using measured data and used for further analysis [6]. The investigation of Botz et al. shows stabilization diagrams of different states of the turbine during a maintenance event. Stable poles are identified and their deviation during the maintenance event are analyzed. The setup is used for long-term monitoring of the structure and serves many research groups as a data basis for their investigations.

To better understand the movements of the tower, it is useful to be able to specify the displacement over the tower height. This cannot be precisely achieved based on acceleration measurements alone, since the double integration of the measured data represents an initial value problem and offset leads to accumulation of integration error. Values of the unknown function at a given point in the domain can be made available by additional sensors. For future research, RTK module on top of the nacelle can be used to provide an estimation of the displacement of nacelle and can be supplemented by a tilt sensor located near the upper acceleration sensor at 139 m. Knowing the displacement and inclination, the accelerometer can be complemented precisely at this point. Figure 2 illustrates the first four bending mode shapes of the tower, that can be observed using accelerometers. The first bending mode can be determined by one single accelerometer as there is an antinode located at the top of the tower. The second, third and fourth bending modes have nodes around the top location of the tower which makes it difficult to detect them with a single sensor.

3. Data Processing

In order to assess the measurability of damage, it is required to carry out an analysis with a real world load spectrum. Main focus of this section is on the challenges associated with the derivation of a load spectrum and the methods used to address them. A general workflow is developed that allows for data-driven simulation of damage cases. An approach is presented that separates the tower displacement calculation into three distinct components. The basic approach is the calculation of dynamic tower displacement using vibrational data. The application of ultra low frequency vibrational sensors with an ultra low noise density of 0.7 µg/Hz to 1.2 µg/Hz enables the determination of even lower frequency measurements up to the first bending

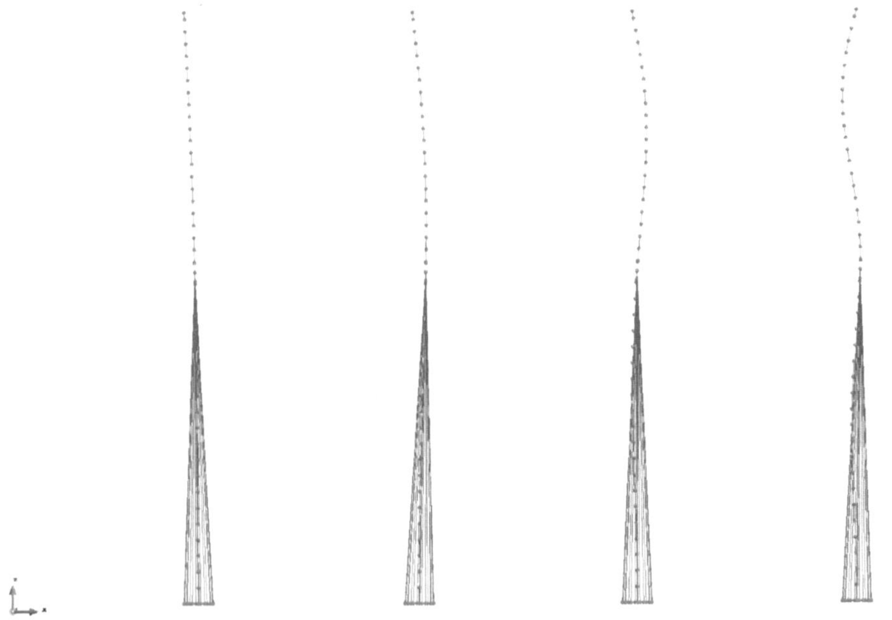

Figure 2. First four bending mode shapes of the tower finite element model.

eigenfrequency at around 0.27 Hz and a further combined natural frequency around 0.23 Hz. Furthermore, the static displacement of the tower, separated into mass induced and thrust force based displacement, is obtained from processed strain measurement data.

Figure 3 depitcs the seperation of static and dynamic displacements at 0.1 Hz which is below all known Eigenfrequencies of this specific wind turbine. To the extent known, below the first natural bending frequency of the tower, only resulting eigenmodes are found, which originate from the rotor speed in operation [6]. As the speed range lies within 7 RPM to 14 RPM, corresponding natural frequencies vary within a range of 0.11 Hz and 0.23 Hz.

In the present research work, the calculation of forces and displacement is based entirely on acceleration and strain measurements and is dealt with in section 3.3 and 3.4. An additional improvement through sensor data fusion is part of future research. An overview of possible combinations is given in section 3.5.

Furthermore, this section covers the required data pre-processing in form of measurement system synchronization in 3.1 and measurement based sensor calibration in 3.2.

3.1. Synchronization

Synchronization of datasets from different sources is an important task for further analysis. Centralized, real-time data gathering with a unique timestamp is always strongly recommended. Application Programming Interfaces (API) of state-of-the-art measurement systems provide for real time data exchange at extremely high sampling rates. If no direct communication is possible, data must be synchronized using alternative techniques. Network Time Protocol (NTP) synchronization is a wide spread and robust method. A further improvement with accuracies up to sub-microsecond range [7] and minimal network and local clock computing resources is the use of Precise Time Protocol (PTP). Independently operating systems require alternative methods such as corresponding measurement signals.

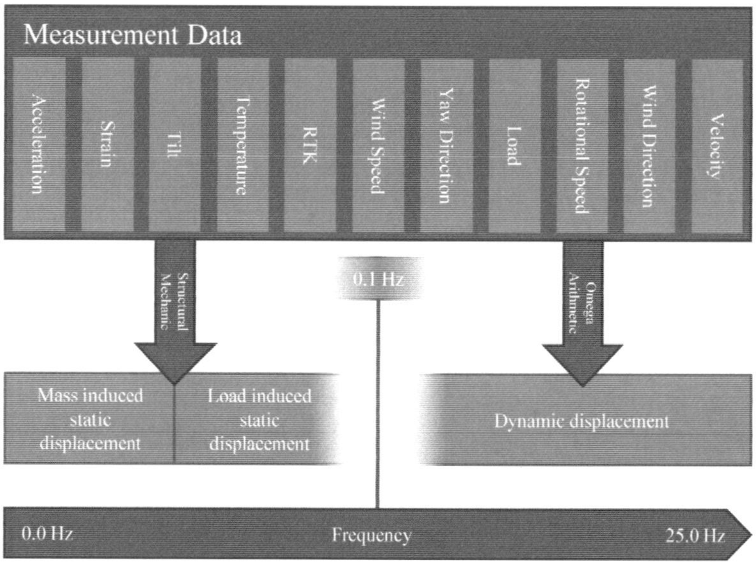

Figure 3. Data processing workflow

Due to the just recently established data interface between SCADA system and the edge controller, historical data requires alternative synchronization, since the timestamps diverge significantly at some points.

Using UCEs with their high repeatability results in precisions ranging up to milliseconds, depending on all other influencing factors. The yaw system of the test facility consists of eight electric gear motors that induce high momentum to supporting structures that further induces high shear stress to the tower due to the large mass moment of intertia of the rotor. Analysis of measurement signals revealed, that shear strain sensors detect a unique pattern of yaw drives starting up. SCADA status messages indicate the exact start of a UCE and are therefore suitable for pattern recognition. A representative pattern, extracted from measurement data is used for recognition using cross-correlation. Due to a duration of the UCE of more than 660 seconds, the extracted pattern is 320 seconds long to ensure viability and precision of detection of start of nacelle movement.

Figure 4 (a) describes yaw direction during an UCE and expected position of extracted and pre-processed shear strain pattern at the beginning of nacelle movement. All strain data are filtered with a high pass filter with a cutoff frequency of 0.1 Hz. Figure 4 (b) shows the acutal position of nacelle movement start in measurement data indicated with a black vertical line. Figure 4 (c) illustrates the aligned position of shear strain pattern due to normalized cross-correlation function shown in figure 4 (d).

3.2. Measurement based sensor calibration and position determination

Because measurement accuracy is highly dependent on the sensor calibration, it is important to ensure proper calibration. A calibration certificate typically specifies sensitivity, frequency response, and offset of a sensor. This section demonstrates, as an example, the measurement-based recalibration of strain sensors using an UCE. The combined rotor and nacelle mass causes a static moment on the tower due to its center of gravity, resulting in strain that rotates with the nacelle position. During a UCE, the pitch angle is approximately 90°, allowing the assumption that the thrust force is low. If this is completely neglected because the wind speed is very low,

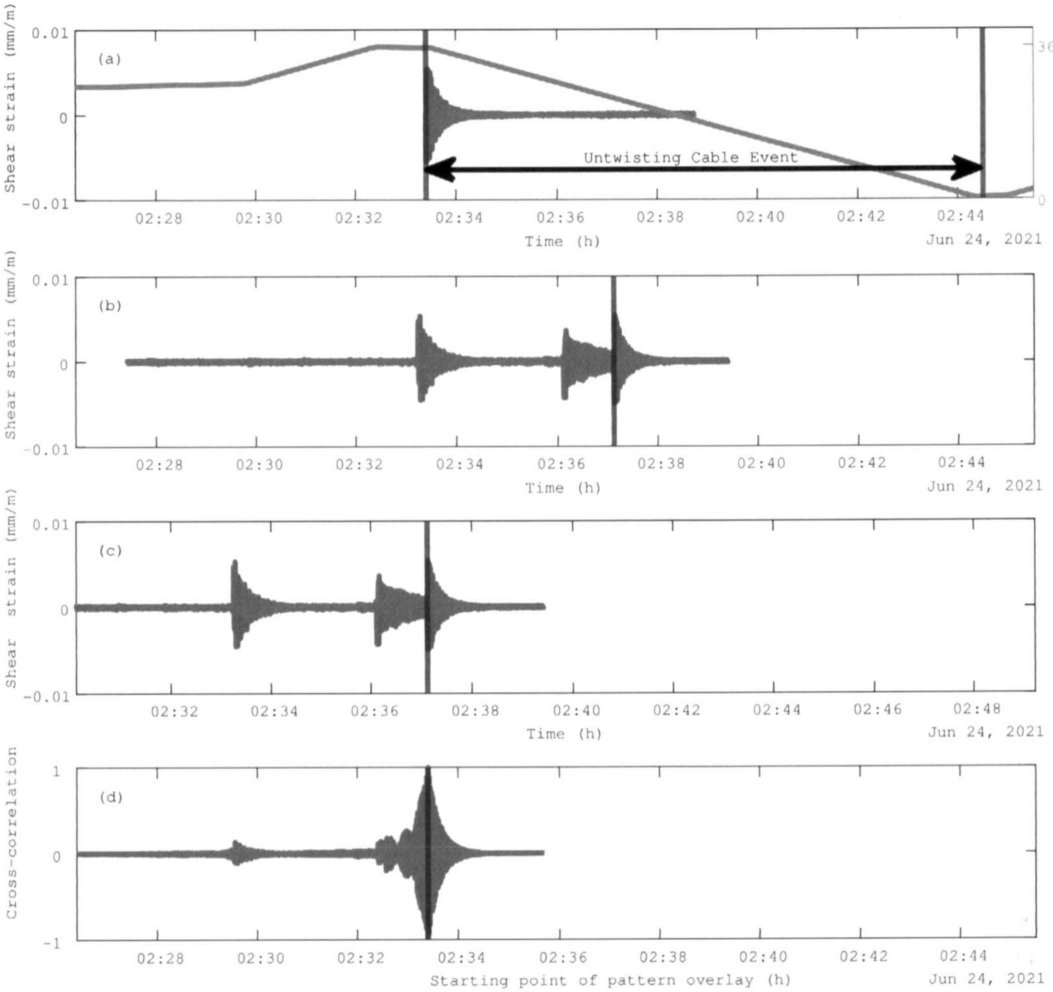

Figure 4. Data synchronization method using shear strain pattern recognition.
(a) Isolated and processed representative shear strain pattern of untwisting cable event (blue) and yaw angle (orange). Arrow indicates duration of UCE.
(b) Highpass filtered shear strain measurement. Horizontal line indicates start of UCE.
(c) Aligned shear strain measurement. Horizontal line indicates start of UCE.
(d) Normalized cross correlation of measurement data and representative pattern.

only static inertia forces impact the tower. The strain profile of a sensor is then represented by a sinusoidal function of the form represented by equation 1. The parameter indicate amplitude ($a = \frac{\epsilon_{max}-\epsilon_{min}}{2}$), period ($b = 1$, as period is equal to 2π), strain gauge position as phase ($c = \alpha$, related to a clockwise rotation from true north in radians) and strain offset ($d = \frac{\epsilon_{max}+\epsilon_{min}}{2}$).

$$f(\theta, a, b, c, d) = a * cos(b * \theta + c) + d \tag{1}$$

Applying curve fitting by means of sum of squared residuals (S) as given in equation 2, leads to strain data during an UCE and thus to parameters a to d by minimizing S as the residual is given by $r_i = f(t) - f(\theta, a, b, c, d)$. Optimization is then performed using a derivative-

Journal of Physics: Conference Series **2151** (2022) 012008 doi:10.1088/1742-6596/2151/1/012008

free optimization algorithm. Figure 5 shows optimized model to an unprocessed real data set. Accuracy satisfies tolerance on function value of $1e^{-6}$.

The applied method can also be used to determine sensitivity, offset and position of accelerometers and tilt sensors as combined rotor and nacelle mass has a great influence on their horizontal axes as the influence of gravity is given by $sin(\alpha) * g$.

$$S = \sum_{i=1}^{n} (\epsilon(t) - f(\theta, a, b, c, d))_i^2 \tag{2}$$

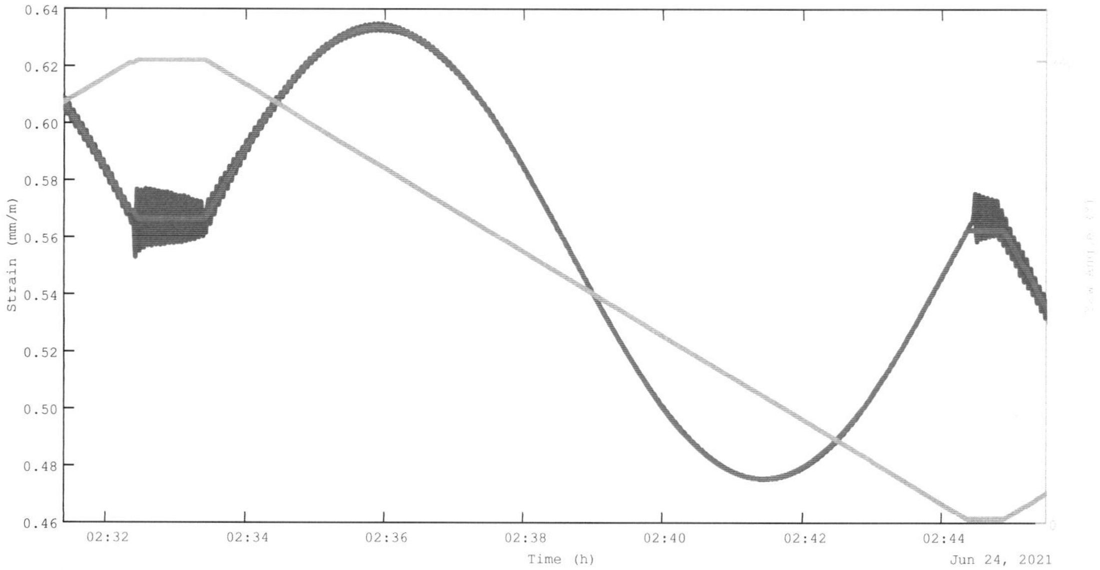

Figure 5. Comparison of unprocessed strain measurement (blue), optimized normal strain model (orange) with parameters $a = 0.079$, $b = 1.000$, $c = 81.053$ and $d = 0.555$ and yaw direction (yellow) during UCE.

3.3. Force calculation

Static forces (< 0.1 Hz) are calculated using strain measurements. The strain is considered separately as the strain due to mass and the strain due to the thrust of the wind turbine. The first can be modeled using the parameterized model from equation 1. Subtracting mass induced strain from measured strain values leads to thrust based strain values. Both static strain components are then used to pure bending moment as $M_b = E * \epsilon_R * S$, where E, ϵ_R and S are Young's modulus, strain and section modulus. Forces are calculated according to $F = \frac{M_b}{h}$, where h represents the distance of strain gauges from hub height.

3.4. Dynamic displacement calculation

Calculation of displacement through acceleration data can be performed time discrete using a cumulative trapezoidal numerical integration or can be performed as frequency domain integration. In terms of real and noisy data, integration in frequency domain tends to be the more precise and reliable way of calculating displacement from acceleration measurement in practical application as stated in [8]. Direct integration in time domain causes drifts due to offset in acceleration or random noise. A method to overcome this issue is proposed by Yang [9] and estimates a mean of upper and lower envelope to correct data.

The present investigation focuses on integration in frequency domain. A general workflow of the application of omega arithmetic, as this method is also referred to, is described in [10]. The practical implementation of the algorithm is inspired by [11].

For the present data set, the frequency range between 0.1 Hz and 25 Hz is considered. Filter parameter of high pass filter to eliminate DC components are set to a stop band frequency of 0.1 Hz and a stop band attenuation of 80 dB. Moreover, a tapered cosine window with a coverage of 4 % of data set is applied.

3.5. Uncertainty quantification

In addition to the systematic and random errors that always appear, the measurements and analysis are subject to specific inaccuracies.

Synchronization of measurement data with SCADA data strongly depends on resolution of the yaw angle. Typically, SCADA data is not subject to a fixed cycle rate, but is event-triggered. As a result, the value is refreshed in response to a specified absolute or relative value change. With respect to synchronization, linear interpolation leads to unacceptable interpolation uncertainty. An interpolation scheme using previous values significantly reduces the error. Taking advantage of the fact that the yaw drives operate at a fixed speed, an advanced interpolation method can be applied. By determining the angular velocity of the nacelle using an UCE, a backwards interpolation with a fixed gradient can be implemented. Figure 6 compares different interpolation methods of low-resolution SCADA data. In the range from 12:47:15 to 12:47:45 it can be observed that both linear and previous value interpolation scheme differs from backwards interpolation scheme. Yaw drives rotate the nacelle constantly from 234° to 250° during this period rather than with interruptions as illustrated by other interpolation methods.

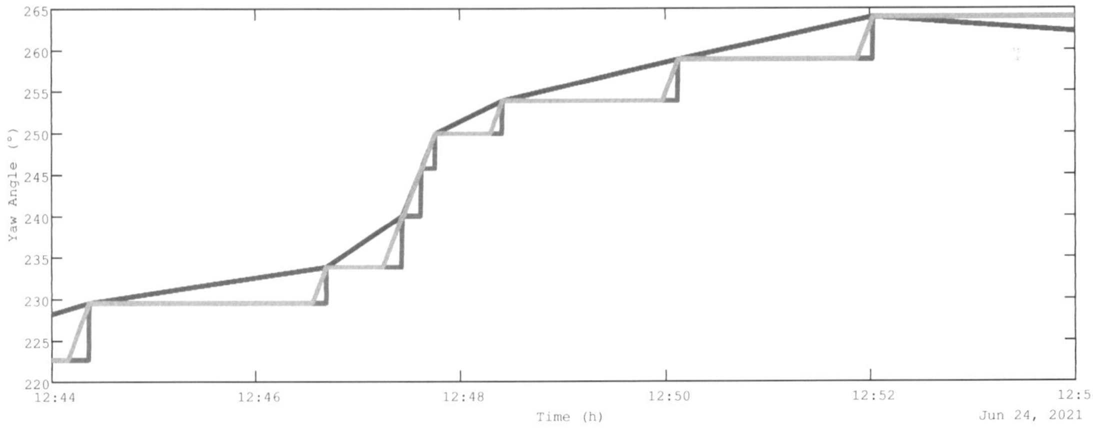

Figure 6. Comparison of untreated and linear interpolated yaw angle from SCADA data (blue), yaw angle using a previous value interpolation scheme (orange) and the developed backwards interpolation scheme with fixed gradient (yellow).

Due to tower bending mode shapes, accelerometer positions are influenced by tilt over time. The influence of tilt is given by $sin(\alpha)$ and is added to the respective horizontal axis. The compensation of these external influences by the respective sensor data alone is a challenging task. Using the Z-axis of the sensor, which represents the vertical axis, the tilt of the sensor can be corrected only to a limited extent, since the change of the z-axis value due to the tilt angle is small $(1 - cos(\alpha))$ and is superimposed by noise. The ratio of the deviations in x/y and z direction is described by $\frac{sin(\alpha)}{1-cos(\alpha)}$. Figure 7 shows both normalized deviations and the resulting ratio. A tilt angle of 0.5° results in a deviation ratio of 229.2. This ratio, being large for small angles, and the superposition by noise do not allow correction of the sensor data by using the corresponding z-axis. Supplementing the influenced acceleration measurement with an inclination sensor, RTK or even both can significantly improve the acceleration and thus calculated displacement results.

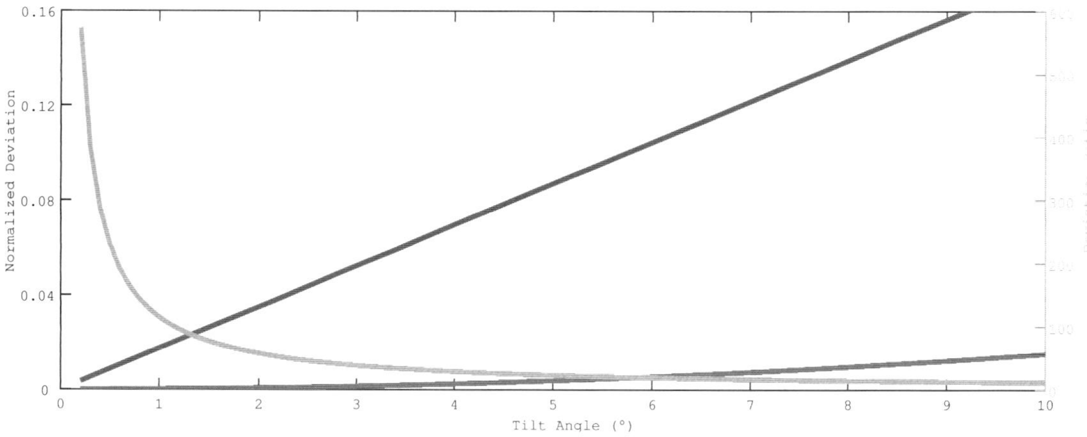

Figure 7. Resulting normalized deviations of x-y-axis (blue), z-axis (orange) due to tilt angle and resulting deviation ratio (yellow).

Furthermore, parametrization of the model developed in section 3.3 for mass induced strain determination is only as precise as wind influences can be excluded. Due to the applied curve fitting algorithm, short wind gusts only have a negligible influence. However, a constant wind has an effect on the parameter and particularly on d, i.e., the sensor offset, and thus influences the absolute calculated values. The number of usable UCEs is noticeably reduced by the fact that rotating winds triggering such an event are often associated with high air flow rates.

4. Simulation Model

The simulation model is achieved by first constructing a CAD model using linear geometrical entities for the simple representation of the tower geometry. Moreover, a Finite Element (FE) model is generated out of the prepared CAD model. The FE model contains cable elements that represent the tendons and beam elements for the modeling of the concrete and steel sections. In particular, one element per concrete and steel section is used. In addition, in order to represent the mass properties accurately, the material density at certain regions is adjusted, such as adapter piece, between the concrete and steel shaft as well as the rotor connection. This model serves as a simplification to the more detailed model that is created using shell and volume elements. The elasticity moduli of the individual reinforced concrete sections are determined by making use of the measured concrete dynamic elasticity modulus [12] and the rule of mixtures [13] in order to achieve a realistic material behavior and incorporate the effect of reinforcement steel.

Table 1. Trend of the first bending Eigenfrequency over fatigue lifetime (n) as a function of $S_{c,max} = [0.9, 0.5, 0.1]$ and $S_0 = 100\%$.

n	0.00	0.25	0.50	0.75	1.00
$f_{S_{c,max}=0.9}$	0.27477	0.27473	0.27416	0.27165	0.26452
$f_{S_{c,max}=0.5}$	0.27477	0.27477	0.27476	0.27390	0.20714
$f_{S_{c,max}=0.1}$	0.27477	0.27477	0.27477	0.27468	0.07748

A particular detail of the Eigenfrequency analysis is the incorporation of the prestresses that occur due to initial displacements that are resultants of self-weight and prestressing tendons. This can be achieved by computing the equilibrium state of the structure under the loading conditions and making use of the stiffness properties at the equilibrium state. The investigations on the initial displacement effects is investigated by Wüchner et al. in [14].

5. Failure Cases

Simulation of the two failure cases missing prestress of tendons *(case 1)* and degradation of reinforced concrete *(case 2)* is expected to result in a direction-dependent change in stiffness and damping of the whole structure. Eigenfrequencies, nodes and antinodes are expected to vary under new conditions.

A simulation with a real load spectrum is performed. The impact on the tower is recorded by measurement. The model is optimized by means of virtual sensors. After adjusting the parameters, simulation is carried out with manipulation of the tower to generate data of a damage case by means of virtual sensors.

Making use of the FE model, a selection of degradation and failure cases are considered and the eigenfrequencies of the tower structure are computed. The considered concrete material degradation cases can be listed as follows; relative fatigue lifetime $n = [0.25, 0.5, 0.75, 1.0]$, fatigue related maximum compressive stresses $S_{c,max} = [0.9, 0.5, 0.1]$. The effect of n as well as $S_{c,max}$ on the concrete elasticity modulus can be calculated using the relations from 3 [15].

$$\begin{aligned} E_c(n) &= E_{c28} * (1 - D_c(n)) \\ D_c(n) &= n^{a_c} * (1 - S_{c,max}) \\ a_c &= 26.5 - 25.0 * S_{c,max} \end{aligned} \tag{3}$$

where $E_c(n)$ is the elasticity modulus of the concrete at relative fatigue life time n, $D_c(n)$ is the concrete degradation coefficient and a_c is the exponent factor . In addition to the material degradation, the relaxation of the prestressing tendons are considered for four different cases; $S_0 = [100\%, 95\%, 90\%, 85\%]$. Table 1 shows that the initial bending eigenfrequency changes significantly towards the end of the service life. The effect is further intensified as $S_{c,max}$ decreases.

6. Conclusion

The enhanced sensor configuration can be used to determine natural frequencies by means of operational modal analysis, to determine stress states in real time and provides knowledge of direction dependent condition data of the whole structure and is thus qualified to perform a comprehensive condition monitoring for supporting structures of wind turbines.

The developed processing workflow provides direct simulation input on two different ways. On the one hand, the input can be provided by calculated static and dynamic forces at the tower top, on the other hand, the direct input can be provided by calculated displacements. The simulation input can thus be generated based on two different measurement techniques. Comparison of both inputs serves as a validation of calculations, leads to a reconciliation of both methods and enables the continuous improvement of the techniques. Operational modal analysis revealed, that acceleration based eigenfrequencies and thrust force based simulated eigenfrequencies match each other.

Comparing vibration-based displacement, processed RTK data, and simulation data with strain measurement-based external force input, the results are remarkably reproducible. The simulation was performed with the updated beam model considering only the first bending natural frequency in two spatial directions. Figure 8 shows a period of time in which the turbine was switched off at medium wind speeds of approximately 4.4 m/s^2 and a wind direction of 315° (NW). Both acceleration and strain data indicate that only the first bending mode is excited during this period. Since the simulation time spanned a period of only 60 s, a settling time of 6 s is excluded from the analysis. The amplitudes and phases of the three independently calculated time series in Figure 8 (a) as well as the natural frequencies of Figure 8 (b) show broad agreement, so the described procedure is accurate enough and can be used for further studies.

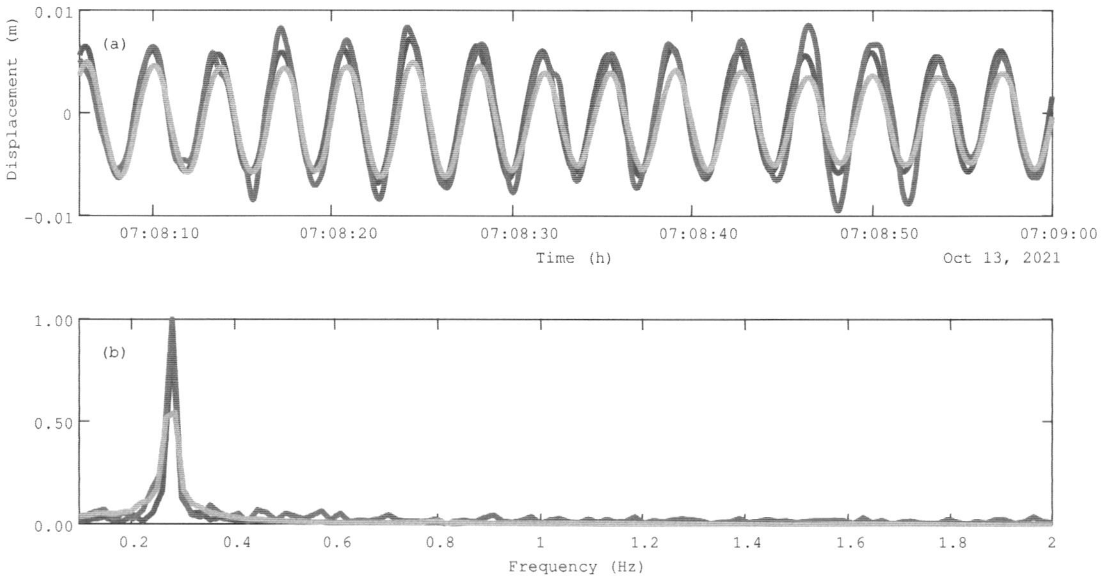

Figure 8. Comparison of dynamic displacement calculated using the three different approaches of vibration based displacement in frequency domain (blue), processed RTK data (orange), and external forces based simulated displacement (yellow)
(a) Comparison of time series.
(b) Normalized frequency spectra of the data sets with peaks at first bending Eigenfrequency of approximately 0.27 Hz.

Directional dependent data for stiffness, natural frequencies and damping can be used to determine pretensioning force of tendons and degradation of reinforced concrete and is able to extend regular maintenance intervals. This results in the capability to perform predictive maintenance, thus reducing downtime and increasing overall equipment effectiveness. A reduced sensor configuration can be used in future systems of condition monitoring for wind turbines to detect further damage and fatigue as well as to determine remaining service life which is part of further investigations within the project.

Acknowledgments

The authors would like to thank the German Federal Ministry for Economic Affairs and Energy for funding the project and Max Boegl and Siemens for providing access to the test facility and graduate student Alejandro Ramírez Piñero for data treatment and assistance during installation. The authors also gratefully thank the staff of Max Boegl for their support during installation of the measurement systems. Supported by TUM International Graduate School of Science and Engineering (IGSSE).

References

[1] Botz M, Emiroglu A, Osterminski K, Raith M, Wüchner R, Große C. Überwachung und Modellierung der Tragstruktur von Windenergieanlagen. Beton- und Stahlbetonbau. 2020 mar;115(5):342-54.
[2] Bellmer H. Schäden an Tragstrukturen für Windenergieanlagen. Schadenfreies Bauen. 2016.
[3] WindEnergie B. Umgang mit Schäden an Fundamenten von Windenergieanlagen. BWE Ratgeber. 2013.
[4] Musiał M, Grosel J. Determining the Young's modulus of concrete by measuring the eigenfrequencies of concrete and reinforced concrete beams. Construction and Building Materials. 2016 sep;121:44-52.
[5] Owen JS. Monitoring the Degradation of Reinforced Concrete Beams under Cyclic Loading. Key Engineering Materials. 2003 jul;245-246:307-14.
[6] Botz M, Zhang Y, Raith M, Pinkert K. Operational modal analysis of a wind turbine during installation of rotor and generator. In: 7th International Operational Modal Analysis Conference (IOMAC); 2017. .
[7] Astarloa A, Rodriguez M, Duran F, Jimenez J, Lazaro J. Synchronizing NTP Referenced SCADA Systems Interconnected by High-availability Networks. In: 2020 XXXV Conference on Design of Circuits and Integrated Systems (DCIS). IEEE; 2020. .
[8] Qihe L. Integration of Vibration Acceleration Signal Based on LabVIEW. Journal of Physics: Conference Series. 2019 nov;1345:042067.
[9] Yanli Yang DK Yanfei Zhao. Integration on acceleration signals by adjusting with envelopes. Journal of Measurements in Engineering. 2016;Vol. 4(Issue 2):p. 117-21.
[10] Jablonski A. Condition Monitoring Algorithms in MATLAB®. Springer International Publishing; 2021.
[11] E Cheynet. Converting acceleration to displacements records. Zenodo; 2020.
[12] Wondra B, Große CU, Raith M, Mayer T, Botz M. MISTRALWIND - Monitoring and inspection of structures at large wind turbines; Teilvorhaben Lehrstuhl für Zerstörungsfreie Prüfung, Arbeitspaket 4: Überwachungstopologie : Schlussbericht BMWi-Verbundprojekt : Laufzeit: 01.08.2015-31.07.2018. Report. 2018.
[13] Tam DKY, Ruan S, Gao P, Yu T. High-performance ballistic protection using polymer nanocomposites. In: Advances in Military Textiles and Personal Equipment. Elsevier; 2012. p. 213-37.
[14] Wüchner R, Osterminski K, Gehlen C, Emiroglu A, Bletzinger KU. Monitoring and inspection of structures at large wind turbines - MISTRALWIND, Teilvorhaben 5 - Lebensdauerakte Windenergieanlagentürme : Schlussbericht : Förderzeitraum 2015-2018. Report. 2019.
[15] Emiroglu A. Multiphysics simulation and CAD integrated shape optimization in fluid-structure interaction [Dissertation]. München: Technische Universität München; 2019.

How do NEWA and ERA5 compare for assessing offshore wind resources and wind farm siting conditions?

P J Meyer and J Gottschall

Fraunhofer Institute for Wind Energy Systems IWES, Bremerhaven, Germany

E-mail: paul.meyer@iwes.fraunhofer.de

Abstract. In advance to the construction of a wind farm, a wind resource assessment is performed, often by the use of wind atlas data. Two widely used and publicly available numerical datasets, ERA5 and the recently created NEWA, are compared and evaluated regarding wind climatology, variability and extreme wind speeds using a comprehensive database with high quality offshore measurements. NEWA shows to be more accurate but less precise in terms of predicting mean wind speed and correlation coefficients. However, the higher temporal resolution of NEWA causes a better representation of the wind variability for the offshore sites. Extreme wind speed predictions based on both numerical datasets are underestimated on average. The spectral correction method is applied and found to be appropriate for the correction of the error in predicting the 50-year return wind speed from the numerical datasets for some sites while it fails for others.

1. Introduction

Due to the high variability of the wind speed, the available wind resource must be examined before any planning or construction of a wind farm. Variations in the wind speed lead to variations in the produced power, which must be known for the economic evaluation of a certain site. Therefore, the wind speed often is measured at the site of interest. However, long-time measurements are scarce, particularly at offshore sites. Wind atlases and numerical datasets are frequently used for the initial estimation of the wind resource at sites of interest, i.e., they serve as long-term reference for the Measure-Correlate-Predict (MCP) method applied within the wind resource assessment [1]. Various numerical datasets with different temporal, horizontal and vertical resolutions exist, each with inherent benefits over the others. A reanalysis dataset widely used in wind energy industry is ERA5, which was proven to be most suitable for wind energy applications among other datasets [2]. However, due to its coarse resolution it fails to predict the wind resource at sites with high topographic complexity [3]. Consequently, mesoscale simulations are performed to tackle this issue. One high-resolution and publicly available mesoscale dataset is the recently created New European Wind Atlas (NEWA) which is based on ERA5 [4, 5].

This study aims to examine how ERA5 and NEWA compare at offshore sites and consequently, to highlight the particular advantages as well as the disadvantages. Since other datasets exists, which can be used alternatively, we want to gain a deeper understanding of the benefits of NEWA. Additionally, a focus is laid on extreme wind speeds and their representation in the numerical datasets. Therefore, a comprehensive offshore measurement database has been created to serve as reference for the comparison. Solely publicly available measurements are considered, such as but not limited to the widely used research station in North and Baltic Sea (FINO) measurements.

Previous research already compared and verified various numerical datasets with measurement data, resulting in different findings. The authors of [6] analyzed and compared NEWA, ERA5 and the Dutch Offshore Wind Atlas (DOWA) for wind characteristics and anomalous wind events for the site of the meteorological (met) mast Ijmuiden and found that each dataset has its advantages. Various numerical datasets, including NEWA and ERA5, were compared in [7] at eight sites in France, finding underestimation of wind speeds by ERA5 in mountainous areas, which was also found by [3], who analyzed ERA5 for various sites worldwide with different topographic and meteorological conditions. Four Lidar measurements in the Baltic Sea were compared to numerical data in [8], and an underestimation of the mean wind speed was found for all datasets with ERA5 showing the lowest errors. The authors in [4] verified and compared the NEWA data with ERA5 regarding wind speed bias and power bias at various sites in Europe and different surrounding terrain, outlining the advantage of NEWA at predicting the mean wind speed.

In the present study, as a differentiation from the other mentioned references, we consider solely offshore sites in Europe to neglect the influence of possible terrain effects but also to address the offshore wind community explicitly. In addition to the wind climatology and variability, we focus on the extreme wind speed events with low recurrence probabilities and their representation within the numerical datasets. Those events need to be considered within the design routine of the wind turbine [9]. Particularly the 10 min-averaged 50-year return wind speed U_{50} with the average occurrence probability of $^1/_{50}$ each year is of importance for the calculation of extreme loads following the guidelines. Further, the spectral correction method (SCM) is applied to the numerical data to correct the loss of the wind variability as consequence of smoothing effects in the numerical models.

2. Materials and methods

2.1. Numerical datasets

Two publicly available numerical datasets are evaluated and compared against each other in this study: NEWA and ERA5. The latter is a reanalysis dataset and widely used in the wind energy community, further it has served as initialization for the generation of NEWA. ERA5 provides values for the wind speed and direction at heights of 10 and 100 m above ground level (agl), though further heights can be accessed using the pressure levels of the model [8]. The spatial resolution is $0.25° \times 0.25°$, corresponding to approximately $28\,\text{km} \times 16\,\text{km}$ at the site of FINO1. The temporal resolution is 1 h. In order to create a higher resolved wind atlas which models mesoscale effects more accurately, NEWA was generated using the Weather Research and Forecasting (WRF) model with enormously high computational effort [4]. It covers all European Union member states and some additional countries on a $3\,\text{km} \times 3\,\text{km}$ grid and provides wind speed and direction values at eight heights ranging from 10 to 500 m agl as instantaneous values every 30 min. Even a microscale atlas, involving further downscaling of the mesoscale data, can be accessed [4] but as we only analyzed offshore sites with negligible terrain effects the mesoscale model was used here. Both datasets are available from 1979 until today (ERA5) and the end of 2018 (NEWA), respectively – more detailed information can be found in [4].

As the temporal and horizontal resolution and consequently the grid box size is significantly smaller for the NEWA dataset, in comparison to the ERA5 data, a smaller distance between measurement site and used grid point can be found on average. An example is shown in Fig. 1 for the site of the Shell Flats (SF) met mast 1 in the Irish Sea. The NEWA grid is much denser and thus the grid box significantly smaller. Particularly onshore and in coastal regions, this can lead to major differences in the terrain character of the grid box. Nevertheless, the direct comparison to observations turns out to be difficult due to the fact that the numerical datasets give instantaneous values averaged for the grid box, while observations describe point measurements averaged for a certain period. Consequently, care must be taken when the numerical datasets are compared to measurement data, which will be elaborated in section 2.3.

2.2. Offshore measurement database

In order to compare and evaluate the aforementioned numerical datasets, corresponding measurements are necessary as reference. Therefore, we created a comprehensive database of wind measurements at offshore locations. Additionally, only publicly available measurements at wind energy relevant heights were considered and the measurements closest to 100 m height were used, as it was done for ERA5 and NEWA. The measurement data was retrieved from four online databases [10, 11, 12, 13]. Furthermore, only the measured time series with a duration of at least 3 years were considered. The database consists of overall 17 measurements from met masts and fixed LiDAR devices located in four countries, their measurement heights ranging from 60 m to 116 m above mean sea level. The sites of the measurements are shown in Fig. 2, together with the wind roses for selected sites, the site name abbreviations can be found in Fig. 3. Note that the sites MMIJ and LIJ (7 and 8) are at the same position in the map, the met mast IJmuiden. All measurements are given as 10 min average values.

After the collection of the measurements, the datasets were quality checked and erroneous values were removed. Some databases already provide quality checked datasets, like the widely used FINO database, while for other datasets only limited documentation and information has been available. We filtered the data for wind speed ($0\,\mathrm{m\,s^{-1}}$ to $60\,\mathrm{m\,s^{-1}}$) and direction ($0°$ to $360°$) range exceedance, repeated values (maximum of two or three consecutive values, depending on the digits) and consecutive wind speed changes of more than $15\,\mathrm{m\,s^{-1}}$ within the 10 min period. Thereafter we combined the met mast wind speed measurements, if measured at booms pointing in different directions at the same height, following the principles in [14] and [15] or used the already quality checked datasets if available. The measurements at a height closest to 100 m was chosen, which is for the met masts mostly the top anemometer. The availability, defined as the ratio of the available measurement points to the maximal possibly available measurement points within a three-day period ($=3 \cdot 24 \cdot 6$ datapoints for 10 min-averaged measurements), of the final datasets are presented in Fig. 3. Note, that the NEWA dataset is available until the end of 2018 only. Supplementary and more detailed information can be found on in [16].

2.3. Wind statistics

In this section, the statistics and methods for the evaluation and comparison of the numerical datasets in this study will be explained. First of all, the ERA5 and NEWA time series for the grid point closest to the measurement site were extracted for each site in the database. Afterwards, we merged the simulation and measurement data in order to yield three datasets with a temporal resolution δ_T of one hour, as this is the lowest resolution of the available datasets (defined by ERA5). However, as mentioned before, care should be taken when comparing grid averaged instantaneous numerical data to point measurements averaged over time, which is 10 min for the measurements. Therefore, multiple approaches for the resampling of measurement data to the temporal resolution of the numeric datasets can be used. While the authors in [5] selected one 10 min period with the least time gap to minute 00, others average the periods to hourly values [6, 7]. For the averaging method, we used a moving average with the timestamp in the middle of the averaging period, i.e., at minute 00. A lower threshold of 4 values per averaging period was chosen for the hourly averaging. For NEWA also a moving average was chosen, using 3 values, so that the values at minute 30 are used for two hourly averaged values. In order to evaluate the two methods, we analyzed the effect of the averaging and selecting on the Power Spectral Density (PSD) estimate. Therefore, we calculated the PSD estimate for the

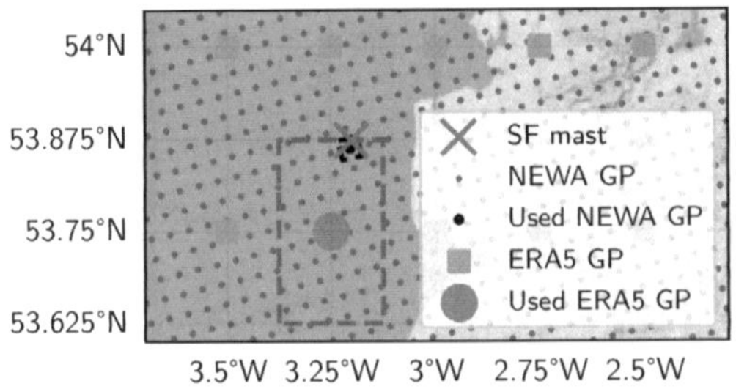

Figure 1. Gridpoints (GP) and met mast at the site Shell Flats (SF)

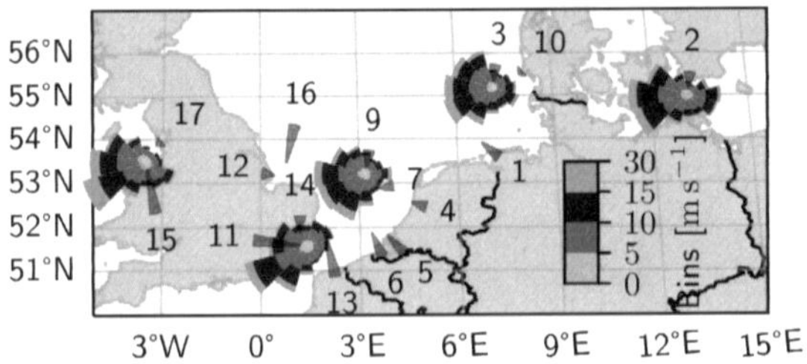

Figure 2. Locations of offshore measurements used in this study, and wind roses for the sites FINO2 and 3, K13a, LA and GYM.

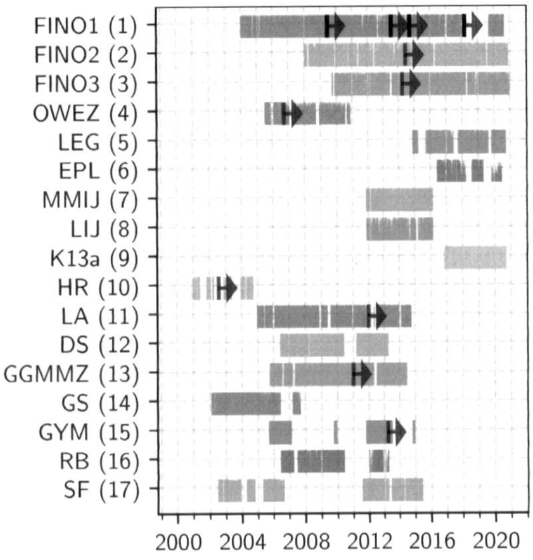

Figure 3. Availabilities of the measurements at all sites. The numbers correspond to the locations in Fig. 2. The arrows indicate the approximate start of wake effects due to adjacent wind farms, which may limit the usability of these datasets.

original measurements with $\delta_T = 10\,\text{min}$ and hourly averaged and selected data as well as for the numerical data by the use of the Scipy package in python [17]. A Hanning window with constant detrending is applied in combination with a window length of 30 days. The exemplary results for the site London Array can be seen in Fig. 4. Looking at the hourly selected measurement, an increase of the PSD can be found within the mesoscale frequency range, while the hourly averaged dataset shows a significantly lower deviation of the PSD estimate within this frequency range. As also the numerical datasets show a loss of the PSD and the increase of the PSD for the selecting method cannot be explained by us, it was decided to apply the averaging method to the measurements for the further analysis. Only data points, where measurements were available, were considered in the analysis. Additionally, the datasets have been filtered for time and wind directions with wakes from wind farms with a distance of less than 10 km to the measurement site. The approximate start times of the wakes are indicated by the arrows in Fig. 3.

After the datasets were prepared, the wind climatology and variability as well as representation of extreme winds in the numerical datasets have been analyzed and evaluated. Therefore, various wind statistics have been used. The mean wind speed $\overline{u}$ is one of the most important parameter within the Wind Resource Assessment (WRA), indicating the wind energy extraction potential of a certain site. Therefore, the bias between the numerical dataset and the measurement, defined as:

$$BIAS_{\text{WS}} = \overline{u}_{\text{sim}} - \overline{u}_{\text{meas}} \tag{1}$$

was used here to illustrate the error, with $\overline{u}_{\text{sim}}$, $\overline{u}_{\text{meas}}$ as mean wind speed from the numerical dataset and measurement respectively. In order to evaluate the suitability of the numerical datasets for the MCP method, the Pearson-r correlation coefficient was calculated and compared, as well as the linear correlation coefficients. For the correlation we only considered hourly values with measured wind speeds of more than $2\,\text{m}\,\text{s}^{-1}$.

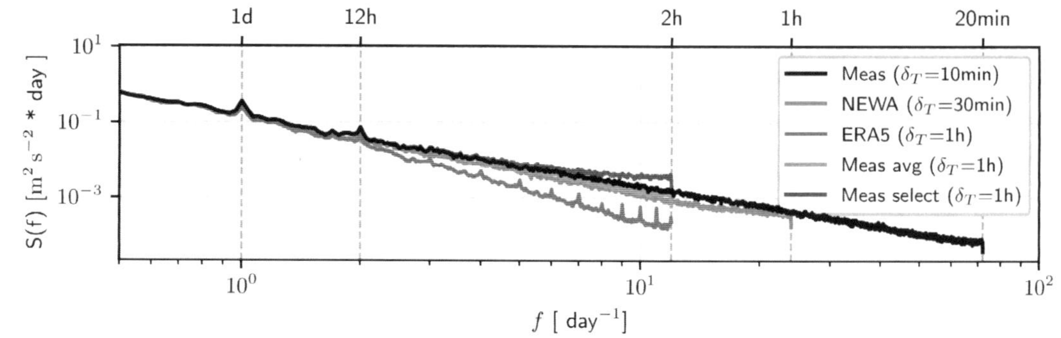

Figure 4. Power Spectral density from measurements and numerical data for the site London Array. The temporal resolution δ_T for each dataset is given in the legend.

Furthermore, also the wind variability is important for the site characteristics. Major variations of the wind speed within short time periods may lead to strong power fluctuations. Consequently, the frequency distribution of the wind speed step changes is analyzed, defined as $\Delta U_i = U_{i+1} - U_i$. Therefore, the original datasets with the temporal resolution of δ_T =10 min, 30 min and 60 min have been used for the measurement, NEWA and ERA5 data, respectively. Here also only data points, where measurements were available, were considered. Additionally, the PSD from measurement and numerical data was analyzed, its calculation has been described before.

2.4. Extreme wind extrapolation

As mentioned before, we laid an additional focus on the extreme wind speed events. Due to the limited lengths of existing wind speed measurement datasets, the extreme wind speeds are often extrapolated from datasets lasting only few years. This is done by the extraction of extreme events and subsequent fitting to extreme value distribution functions. Various methods and distribution functions for the extrapolation exist and are used in the wind energy industry, an overview can be found in [18, 19].

Within this study, the peak-over-threshold (POT) method is preferred over the widely used annual maxima method (AMM) [20, 21] because of the required length of the underlying dataset. While for the AMM at least 10 years of data is needed, for the POT method a minimum of 5 years of data is recommended [18]. The method requires the determination of two variables, the lower threshold value u_{thresh} and the storm separation time Δt that must be between two consecutive peaks. We used a storm separation time of 7 days, which was shown to be a reasonable time for the analyzed sites in [19]. The wind speed threshold value however has no physical derivation, while its value has a major influence on the extrapolation result. Its determination may be difficult, yet a value about the wind speed of the lowest annual maximum within the dataset was shown to be reasonable [19]. Additionally, the shape factor of the underlying distribution function was set to zero. Generally, larger sample sizes lead to lower standard errors of the extrapolation. Therefore, we only considered the datasets with a length of more than 5 years and a high availability without major gaps for the evaluation of the extreme wind speeds. Further, only years, defined as the 12-month period starting from the 1th of July, with a data availability of more than 50 % were considered for the extrapolation. Here, only periods, where at least two measurement points per hour are available, have been used, to prevent discrepancies due to missing measurements. Also, the original temporal resolution is used, as the parameter of interest U_{50} is defined as 10 min averaged value.

However, smoothing effects and the limited temporal resolution of the numerical datasets lead to lower estimates of U_{50}. Consequently, the SCM was used here to take into account the missing variability, which was shown by the PSD estimates in Fig. 4. The method was first developed in [20] and further adapted and improved in subsequent studies, such as [21]. Firstly, the PSD from the numerical dataset is corrected for $f > f_c$ by measurements, or as done here with a function of $S(f) = a \cdot f^{-5/3}$, which was shown to be the slope of the measured spectrum in the mesoscale range [20]. Afterwards, this spectrum is extended to the frequency of $72\,\mathrm{d}^{-1}$, which corresponds to the Nyquist frequency of the 10 min sampling frequency. The U_1 extreme wind speed with a yearly average recurrence period can then be calculated by various mathematical assumptions and the use of the spectral moments from the original and corrected spectrum, see [20, 21]. The ratio of the corrected to the uncorrected U_1 is then defined as the correction factor CF, describing the general ratio of a corrected and uncorrected extreme wind speed U_T and thus is also valid for U_{50} [21].

For an optimal performance of the method, the calculated correction factor should correspond to the ratio of the U_{50} value from the measurements to the U_{50} from the numerical dataset. However, the frequency f_c, which must be defined, has a major influence on the results. The authors of [20] used $f_c = 2\,\mathrm{d}^{-1}$ for most sites, while the authors of [21] used $1.5\,\mathrm{d}^{-1}$ which also fits well for the sites in this study and consequently was adopted here. Therewith, we can calculate the correction factors for the NEWA and ERA5 datasets at each site. The Authors of [21] observed the upward bend of the power spectrum from the NEWA dataset which might be caused by the high temporal resolution and thus non-physical but artificial noise within this dataset. Consequently, they proposed to use a two-point running average, leading to a major reduction of the spectral energy in the mesoscale frequency range. This method is called SCM NEWA Bastine in the following. Additionally, the hourly averaged NEWA dataset using a three-point average, as explained in section 2.3 is shown as reference, named SCM NEWA 1h. Finally, also the correction factor for the available 30 years of data from NEWA and ERA5 have been calculated to evaluate the influence of the length and availability of the reference measurement dataset on the correction factor, called SCM NEWA 30y and ERA5 30y.

3. Results

3.1. Wind climatology

The most important parameters for the WRA, the $BIAS_{\mathrm{WS}}$ and the correlation coefficient r for all sites and both numerical datasets are shown in Fig. 5. On average, both numerical datasets underestimate the mean wind speed by up to $0.5\,\mathrm{m\,s^{-1}}$ and $1\,\mathrm{m\,s^{-1}}$ (NEWA and ERA5, respectively). On the one hand, the absolute bias is on average lower for the NEWA dataset, with the mean bias being only $-0.19\,\mathrm{m\,s^{-1}}$, indicating a higher accuracy compared to ERA5 with a mean bias of $-0.36\,\mathrm{m\,s^{-1}}$. Also, the spread of the bias, indicated by the standard deviation in the legend, is significantly lower for the analyzed sites, when the NEWA dataset is used. For the NEWA dataset, the mean wind speed is overestimated at two sites (GS and GYM), while it is only overestimated at one site (HR) with ERA5. On the other hand, the correlation coefficients are higher for all sites, if ERA5 is used, compared to NEWA, being 0.95 and 0.89 on average, respectively.

3.2. Wind variability

Apart from the wind climatology also the variability has been analyzed for each site individually. Doing so, multiple observations can be made, which are valid for most of the analyzed sites. They are outlined here for the example of London Array. First of all we look at the $0.1\,\mathrm{m\,s^{-1}}$ binned frequency distribution of the step changes for the measurement, NEWA, ERA5 data as well as the hourly averaged and selected datasets in Fig. 6. Despite the fact, that NEWA and the measurements have a different temporal resolution, they show a very similar frequency distribution for wind speed step changes within the range of $\pm 1\,\mathrm{m\,s^{-1}}$, which accounts for the

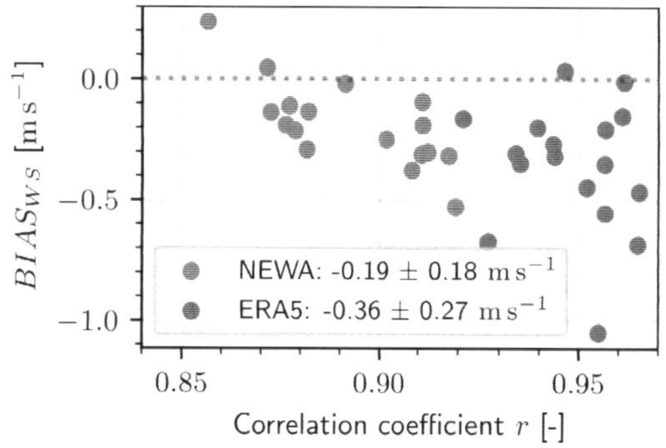

Figure 5. Bias and Correlation coefficient for all sites and both numerical datasets within this study, $BIAS_{\mathrm{WS}}$ mean and one standard deviation are given in the legend.

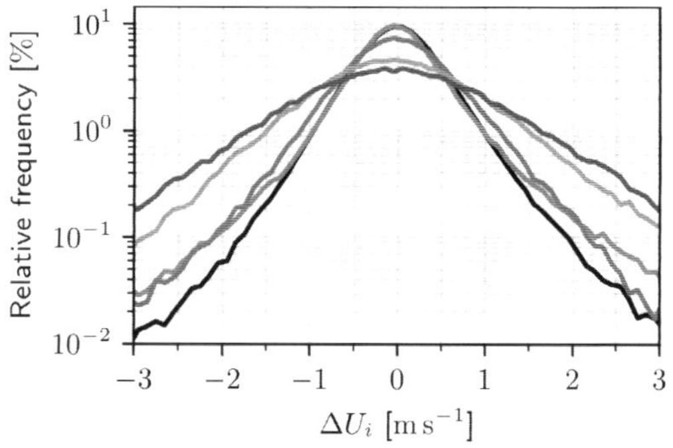

Figure 6. Frequency distribution of Wind speed step changes with at the site London Array. The colors are the same as in Fig. 4. The bin width is $0.1\,\mathrm{m\,s^{-1}}$

majority of the step changes. However, for larger step changes, the frequency is lower for the measurements, which can be traced back to the lower temporal resolution. For the same reason, ERA5 shows a lower frequency for the small step changes. The hourly averaged and selected measurements show significantly lower probabilities for low wind speed step changes and higher probabilities outside the range of $\pm 0.5\,\mathrm{m\,s^{-1}}$, though the averaged distribution is more similar to the original distribution. For some sites with large distance to the shore, such as K13a, FINO3, LIJ, the probability for this range was observed to be even slightly higher for the NEWA dataset compared to the measured distribution. Consequently, the probability for the larger step changes is lower compared to the measurements.

The observations made within the wind speed step analysis match the observations from the PSD estimate, which was shown exemplary for the site London Array in Fig. 4. Apart from the fact, that the peaks at frequencies of $1\,\mathrm{d^{-1}}$ and $2\,\mathrm{d^{-1}}$ are covered by both numerical datasets, they show a loss of the PSD estimate for frequencies higher than these peaks. However, the loss of the power is largest for ERA5, while it is significantly lower for the NEWA dataset. Very similar observations were made for all analyzed offshore sites within this study.

3.3. Extreme wind speed

First of all, the representation of measured extreme wind speed events within the numerical datasets was analyzed and compared. Therefore, we used the 30 highest measured peaks over threshold from all sites with a storm separation time of 7 days. Then we extracted the highest wind speed within half of this storm separation time from both numerical datasets. For each extreme wind speed event the bias is then calculated, as in Eq. 1 but with the peak wind speeds instead of averages. The results can be found in Fig. 7, where a scatter plot of the bias as function of the measured wind speed is shown next to the frequency distribution of the bias. For both numerical datasets, most data points can be found in the range of $-5\,\mathrm{m\,s^{-1}}$ to $1\,\mathrm{m\,s^{-1}}$. The average absolute bias is significantly lower for the NEWA dataset, whereas the spread is larger, indicated by the normal distributions in the right plot. Particularly for larger wind speeds, the underestimation is larger for NEWA, this is the case for all measured wind speeds above $35\,\mathrm{m\,s^{-1}}$. For the ERA5 dataset rarely an overprediction of the extreme wind speeds is observed.

Consequently, the extrapolated wind speed U_{50} is underestimated for most sites if the numerical datasets are used. Therefore, the SCM was applied to the numerical datasets. The results can be seen in Fig. 8, where correction factors and U_{50} ratios are shown. A U_{50} ratio of 1 indicates, that no correction is necessary. Inspecting the diamonds, we could find that the

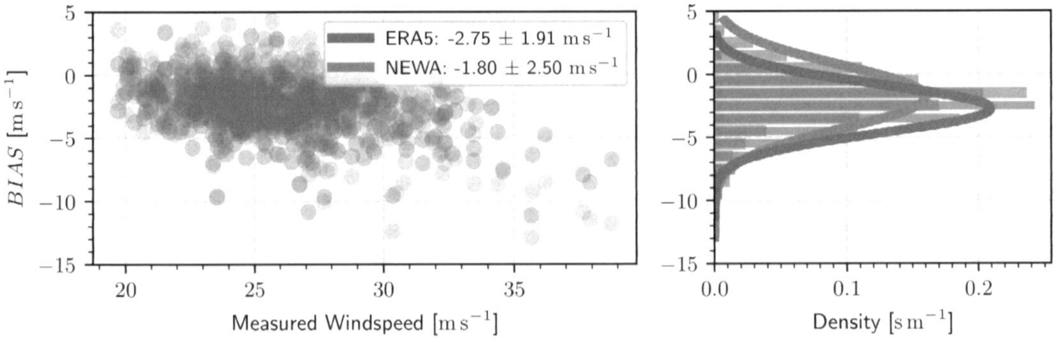

Figure 7. Highest 30 measured winds peeks and the corresponding highest wind speed in the numerical dataset within a storm separation time of 7 days. The mean and standard deviation are given in the legend, corresponding to the normal distribution in the right plot.

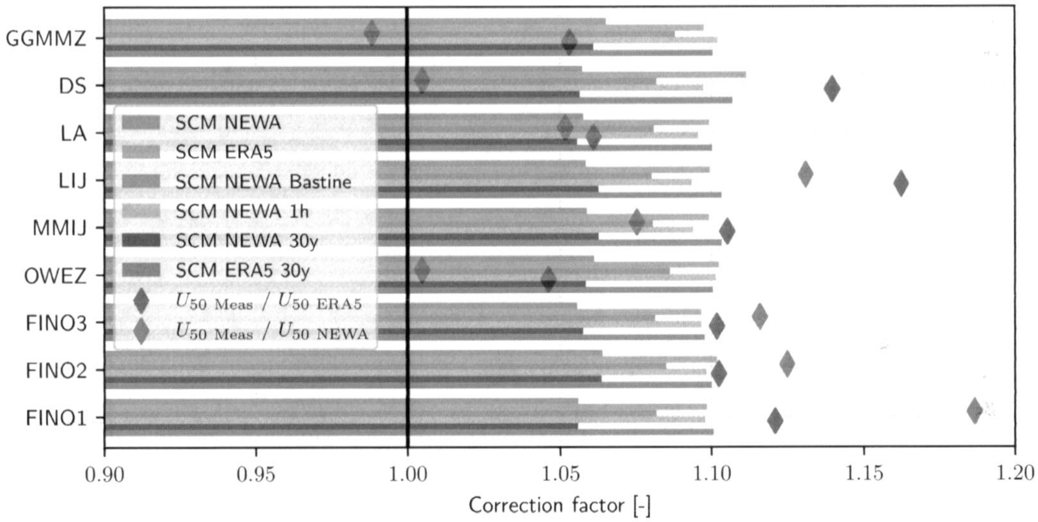

Figure 8. Correction factors for the estimation of U_{50} using the SCM for various datasets, which were explained in section 2.4. Further, the U_{50} ratio from the measurement to the numerical datasets are shown as diamonds.

U_{50} extrapolation, using the POT method, leads to an underestimation for most sites, when the uncorrected numerical datasets are used. Solely for the site GGMMZ an overestimation can be found for NEWA, which might be caused by the short time series available. The uncorrected U_{50} ratios vary within a range of 0.99 to 1.19 and they are smaller for most sites, if NEWA is used. Only for the German FINO sites the U_{50} are extrapolated more accurately by the ERA5 datasets. Yet for the three sites of GGMMZ, DS and OWEZ almost no correction is necessary, if NEWA is used for the extrapolation. Nevertheless, the U_{50} ratios show major variations, and no recognizable correlation was observed. On the other hand, the corresponding correction factors are similar for all sites. The lowest correction factors of approximately 1.06 can be found for the original NEWA dataset. Only small differences were found, when the PSD only from the measurement period or from 30 years is corrected. Using the Bastine method, correction factors of 1.08 to 1.09 were observed. Using the 3-point hourly averaged NEWA datasets, correction factors of 1.09 to 1.10 are calculated, while the highest correction factors of approximately 1.10 can be found for the ERA5 datasets. Nevertheless, the correction factors do not uniformly represent the real U_{50} ratios for the sites in this study. Only for the sites MMIJ, FINO3, FINO2 and approximately FINO1 the correction factors from ERA5 represent the uncorrected U_{50} ratios gratifying, while the U_{50} is then overestimated for the sites GGMMZ, LA and OWEZ. The difference in LIJ and MMIj, despite the same site, is caused by the fact, that the highest measured wind speed were recorded to be higher from the met mast.

4. Discussion

NEWA has been created based on the ERA5 dataset to provide a publicly available high-resolution dataset of wind resource relevant parameters for the European region. However, the creation is associated with high computational effort. Both datasets have been compared and evaluated by the use of a comprehensive measurement database at offshore sites in order to assess and understand the benefits and shortcomings of the new dataset.

The analysis of the wind climatology and the comparison between the two numerical datasets showed, that ERA5 is more precise at all sites in terms of the correlation coefficient and consequently might be the preferred dataset for the long-term MCP method within the WRA. Those observations confirm the findings of [7], who found the highest correlation coefficients among various numerical datasets for ERA5. Further, the correlation coefficients are higher and the bias lower compared to the onshore study in [7]. On the other side, NEWA showed lower bias on average and thus to be more accurate in terms of the mean wind speed for the analyzed offshore sites. This corresponds to the findings in [4], who found that NEWA models the mean wind speed with lower bias than ERA5. Here, the numerical data at a height of 100 m was extracted, whereas the measurements have deviations of up to 40 m in height. Therefore, in future studies, the height shall be considered more accurately, although the effect is expected to leave the results at offshore sites unchanged.

Another advantage of the NEWA dataset is the representation of the variability. The frequency distribution of the wind speed step changes is modeled accurately by NEWA for most sites in this study within the range of $\pm 1\,\mathrm{m\,s^{-1}}$, yet overestimates the probability for large step changes. However, different temporal resolutions are chosen, which makes a direct comparison difficult and further investigation necessary. Both numerical datasets model the PSD accurately up to frequencies of $1\,\mathrm{d^{-1}}$, yet for higher frequencies the spectrum from ERA5 shows a major deviation due to smoothing effects and lower temporal resolution. NEWA shows a higher and more realistic variability of the wind speed at a temporal resolution of 30 min.

The representation of the extreme wind speeds has been analyzed by the extraction of the highest 30 extreme wind speed events from the measurements and their corresponding value in the numerical dataset. Only the measurements with a time period of at least five years were considered to have a reliable basis for the extraction of extreme wind speeds. We can conclude that these events are in general captured and presented in the numerical datasets, yet their magnitude is on average underestimated by both. NEWA shows a lower underestimation, although a negative correlation coefficient is found, meaning the underestimation becomes larger with increasing wind speed. Especially for the highest measured wind speeds in the database, NEWA shows a significantly greater underestimation compared to ERA5. The underestimation of the extreme wind speeds also leads to an underestimation of the extrapolated U_{50} 10 min averaged wind speed, which is part of the guidelines [9].

The reasons for the underestimation are the smoothing of variables for numerical stability in the datasets and the limited temporal resolution. Therefore, the SCM was applied to tackle this problem. However, the correction factor, which is applied to the U_{50} from the numerical datasets, is similar for all sites analyzed in this study. Only little variation can be seen, while the ratios of the U_{50} from measurement to numerical dataset show major variations. For three sites the U_{50} is modeled accurately, when NEWA is used in combination with the POT method and without the SCM. Overall, NEWA predicts the U_{50} with higher accuracy compared to ERA5, except for the FINO sites. However, they are widely used and represent the dataset with the highest quality and length. Therefore, further long-term measurements have to be included in further research. The correction factors do not vary much, as the PSD estimates are similar for the sites. In general, the correction factors are lower for the NEWA dataset, as the temporal resolution is higher and the variability does not deviate from the measurements as much as for ERA5. Further, we chose the parameter f_c to be $1.5\,\mathrm{d}^{-1}$ for all sites, while more advanced methods exist which should be included in further research, the uncertainties have been discussed briefly in [21]. The correction factor approximately fits the ratio of the U_{50} only at few sites. This however might be caused by the limited length and quality of the reference measurement dataset. However, a high quality and availability of the reference dataset is crucial for the evaluation of the extreme wind speeds, as the uncertainty decreases with the length of the dataset.

5. Conclusion

Two widely used numerical datasets, NEWA and ERA5, have been compared and evaluated within this research. Past studies already elaborated the shortcoming of ERA5 for complex onshore terrain. This study however analyzed and highlighted the advantages and disadvantages of NEWA over ERA5 at offshore sites by the use of a comprehensive offshore measurement database. The following results have been found for the offshore sites within this study:

- On average, NEWA models the mean wind speed more accurate than ERA5. On the other hand, a significantly higher correlation coefficient was calculated for all sites ($r > 0.92$), if the ERA5 datasets was used, proving its suitability for the MCP method.
- The variability of the wind speed, in terms of the power spectral density as well as the wind speed step changes, is modeled more accurately by NEWA. ERA5 has a lower temporal and horizontal resolution, leading to a less accurate description of the wind speed variability.
- Measured extreme wind speeds are modeled but underestimated by both numerical datasets. The underestimation on average is lower for NEWA, though the underestimation rises with the wind speed.
- The U_{50} correction factors using the SCM are similar for all offshore sites (1.05 to 1.11, depending on the datasets), while the ratio of over- and underestimation varies between 0.99 to 1.19. For some sites, the correction factor can be applied satisfyingly to the extreme winds, while for other sites this would lead to an overestimation or still underestimation.

Overall, there is an advantage of NEWA over ERA5 in terms of the prediction of mean wind speed and extreme wind speed extrapolation, though the advantage is minor. ERA5 convinces with its suitability for the long-term MCP method. However, in terms of the wind variability, NEWA convinces due to its high temporal resolution and consequently realistic representation of the variability to be used for further time-series-based wind resource assessments.

Acknowledgments

This study is part of the project Digitale Windboje supported by the Federal Ministry for Economic Affairs and Energy of Germany under Grant No. 03EE3024. Further this study has benefited from measurements downloaded from various internet databases, such as "Database of Wind Characteristics" located at DTU, Denmark, the Marine Data Exchange, "Wind op Zee" from TNO Wind Energy, the FINO initiative, which we would like to thank for the access to their databases. Lastly, we appreciate the provision of the numerical datasets by our colleague Dr. Martin Dörenkämper, Fraunhofer IWES.

References

[1] Rohrig, K and others 2019 *Applied Physics Reviews* **6** 031303
[2] Ramon J, Lledó L, Torralba V, Soret A and Doblas-Reyes F J 2019 *Quarterly Journal of the Royal Meteorological Society* **145** 3236–3251 ISSN 0035-9009
[3] Gualtieri G 2021 *Energies* **14** 4169
[4] Dörenkämper *et al.* 2020 *Geoscientific Model Development* **13** 5079–5102
[5] Hahmann A N *et al.* 2020 *Geoscientific Model Development* **13** 5053–5078
[6] Kalverla P C, Holtslag A A M, Ronda R J and Steeneveld G J 2020 *Quarterly Journal of the Royal Meteorological Society* **146** 1498–1515 ISSN 0035-9009
[7] Jourdier B 2020 *Advances in Science and Research* **17** 63–77
[8] Hallgren C, Arnqvist J, Ivanell S, Körnich H, Vakkari V and Sahlée E 2020 *Energies* **13** 3670
[9] IEC 2019 *IEC 61400-1: Wind energy generation systems –Part 1: Design requirements* edition 4.0 ed (Geneva)
[10] FuE Kiel GmbH 2021 Fino1, 2, 3 homepage URL `https://www.fino-offshore.de/de/index.html`
[11] The Crown Estate 2021 Marine data exchange-wind data URL `http://www.marinedataexchange.co.uk/wind-data.aspx`
[12] TNO Wind Energy Wind op zee URL `https://www.windopzee.net/en/`
[13] Larsen G C and Hansen K S winddata database URL `www.winddata.com`
[14] Werkhoven and Verhoef 2012 Offshore meteorological mast ijmuiden: Abstract of instrumentation report: Ecn-wind memo-12-010 Tech. rep. ECN URL `https://www.windopzee.net/wp-content/uploads/2019/07/ecn-wind_memo-12-010_abstract_of_instrumentatierapport_meetmast_ijmuiden.pdf`
[15] Leiding T *et al.* 2016 Abschlussbericht: Standardisierung und vergleichende analyse der meteorologischen fino-messdaten Tech. rep. Förderprogramm des Bundesministeriums URL `https://www.dwd.de/DE/klimaumwelt/klimaforschung/klimaueberwachung/finowind/finodoku/abschlussbericht_pdf.pdf?__blob=publicationFile&v=3`
[16] Meyer P J 2021 *Development of a Methodology for the Verification of Numerical Wind Data* Master thesis University of applied Science Flensburg Flensburg
[17] Virtanen P *et al.* 2020 *Nature Methods* **17** 261–272
[18] Palutikof J P, Brabson B B, Lister D H and Adcock S T 1999 *Meteorological Applications* **6** 119–132 ISSN 13504827
[19] Larsén X G, Mann J, Rathmann O and Jørgensen H E 2015 *Wind Energy* **18** 59–74 ISSN 1095-4244
[20] Larsén X G, Ott S, Badger J, Hahmann A N and Mann J 2012 *Journal of Applied Meteorology and Climatology* **51** 521–533 ISSN 1558-8424
[21] Bastine D, Larsén X, Witha B, Dörenkämper M and Gottschall J 2018 *Journal of Physics: Conference Series* **1102** 012006 ISSN 1742-6588

Developing a long-lasting offshore wind business case towards a Dutch decarbonised energy system by 2050

S Krishna Swamy[1], I Gonzalez-Aparicio[1] and N Chrysochoidis-Antsos[1]

TNO Energy Transition, 1755 LE Petten, The Netherlands
Email: siddharth.krishnaswamy@tno.nl

Abstract The integration of large-scale offshore wind power in the energy system will have a major effect on electricity markets. It may lead to large market price volatility due to the inherent variability of wind energy in terms of power fluctuations, forecast errors and insufficient flexibility on the demand side. This study models future business cases for offshore wind farms in two reference national scenarios for the Netherlands, namely NECP2030 and TRANSFORM. Results show that the value of offshore wind in 2030, according to NECP scenario is 39.9 €/MWh, and that electricity price in the TRANSFORM scenario, which considers aggressive development towards a sustainable economy, is a high 104 €/MWh by 2050. To consider ways of improving the return on investment for offshore wind farms in future markets, an optimisation problem is defined, where the wind farm has the choice to either sell the generated electricity to the spot market or to an electrolyser. Results show a potential advantage in using an electrolyser to produce green hydrogen for offshore wind farms, with net profit of 41.4 M€, compared to when offshore wind farms sell electricity solely to the spot market, where the profits are 36 M€.

1. Introduction & Literature

New obligations laid down in the first European Climate Law advocate for 55% net greenhouse gas (GHG) emission reductions by 2030 (compared to 1990 levels), on the way to European climate neutrality by 2050 [1]. The blueprint for this transformational change puts in place landmark strategies for offshore renewable energy. In particular, offshore wind energy in the European union is to increase from its current capacity of 12 GW to at least 60 GW by 2030 and to 300 GW by 2050. The Netherlands targets an offshore wind portfolio of at least 11.5 GW by 2030 with the vision of reaching 60 GW to 75 GW by 2050 in the Dutch North Sea [2].

Due to their inherent variability, the high share of electricity production from renewable energy sources (RES) increases stochasticity in the electricity market. This could lead to increased risk premiums and without subsidy support schemes, wind farm developers will be exposed to the price volatilities of electricity markets. This imposes increased risks of revenue losses, for instance due to unforeseen price drops, curtailments or inability to adapt to variations in the market and may also give rise to other market opportunities [3].

System integration, with conversion and storage technologies such as green hydrogen production, offers a promising prospect to develop long-lasting business cases for offshore wind farms (OWFs), while also increasing flexibility in the electricity market. However, it is still unclear how this can be achieved without endangering the stability of the network, the technical and economic considerations needed and the conditions that are profitable to OWFs in this integrated system. Energy and power system models, are widely used tools to evaluate future scenarios, yet the emergence of variable RES, hybrid systems and new market mechanisms have required modellers to adapt both the methodology and the datasets used.

Regional cooperation to meet the flexibility challenge posed by high RES penetration in the EU is explored in [4]. An analysis of electricity interconnection potential to make use of synergies between the UK and France is seen in [5]. In Germany, to support the expansion of wind and solar power in the energy market, research into power to gas and power to liquid (P2X) technologies is recommended in [6]. A ten network development plan from the European network of transmission system operators (ENTSO-E) indicated that the cost and availability of flexibility providers such as P2X and batteries will influence investment decisions in RES [7]. Strategies to offset the cannibalisation effect of RES are explored in [8] and combining a RES investment with a flexible power supply contract (e.g. electrolysis) appeared to be a more effective hedging strategy. METIS, an energy modelling software, was used in [8], and can simulate the European energy system for electricity, gas and heat to high spatial and temporal granularity [9].

In comparison to the above literature, this study contributes in the following ways. This study quantifies the effects of integrating large amounts of RES, in particular offshore wind energy, into the Dutch electricity market and its impact on electricity prices. This is done with the use of a newly developed first order estimation tool which simulates the clearing mechanism of national wholesale electricity markets. The business case for upcoming OWFs in these future scenarios is investigated, as is the prospect of converting their energy into hydrogen. In particular, the techno-economic conditions needed for OWFs to obtain higher profits by either selling energy in the electricity market or to produce green hydrogen is looked into.

This paper first introduces the newly developed market modelling tool in section 2. This is followed by considering future scenarios and evaluating the future price of electricity and value of offshore wind energy in section 3. Section 4. describes a potential solution to improve the business case for OWFs with their integration to an electrolyser to produce green hydrogen. Section 5. discusses the conclusions and further work.

2. Methodology

The study models the Dutch electricity market in several scenarios with a special focus on the offshore wind development roadmap [10]. Offshore wind profiles are obtained based on the wind speed timeseries at the center coordinate of the wind farms in 2030. Based on the wind speed timeseries, an energy yield timeseries is obtained based on an interpolation from a sample 1 GW wind farm energy yield rose plot, calculated using TNO's internal wake loss estimation tool ECN Farm Flow [11].

2.1. Market model

The Dutch electricity market is modelled using the EYE (ElectricitY market price Evolution simulator) model [9]. The EYE model is an electricity system simulator which can analyze electricity prices given certain scenario inputs (such as energy asset specifications, commodity prices and expected demand). The EYE model is developed to model the future electricity grid and flexibility options with a first order estimation, in order to study complex system effects quickly.

2.1.1. Merit order: Using a merit order supply bid ladder, the model dispatches the cheapest available energy supply asset that is defined in each simulation step to cover the respective demand from the electricity system. The marginal cost of each supply asset is determined by the asset fuel price and its efficiency using the relation:

$$Marginal\ cost\ \left(in \frac{€}{MWh}\right) = Fuel\ price\ \left(in \frac{€}{MWh}\right) + CO2\ cost\ (in \frac{€}{MWh}) \quad (1)$$

2.1.2. Flexible demand: Besides must-run electricity demand, an advantage of the model is the possibility to simulate flexible assets (such as batteries, electrolyzers, hybrid boilers and industrial heat pumps) which represents the future of electricity market demand. These assets have a number of characteristics which could be set for each simulation (e.g. bidding strategies, forecast window lengths, round trip costs, relevant efficiencies, nominal capacities etc.).

2.1.3. Price clearing: The price clearing is done by finding the intersection of the supply and demand bid ladders. By providing as inputs the characteristics of electricity supply assets (e.g. wind, solar, natural gas and coal), fuel prices, CO_2 taxation, renewable asset production profiles, electricity demand and flexibility assets, the model derives a market clearing price on an hourly basis, thus simulating future spot market prices. With output market clearing prices, the annual value of offshore wind is estimated by the equation below:

$$Value\ of\ offshore\ wind\ (in\ €/MWh) = \sum_{i=1}^{8760} (p_i * c_i) \,/\, \sum_{i=1}^{8760} (p_i) \tag{2}$$

where p_i is the electricity sold from all offshore wind assets and c_i is the resulting market clearing price at each simulation timestep.

3. Scenarios to evaluate future value of OWFs

Offshore wind energy development has traditionally been driven by government support schemes. Recent projects however have tended to move towards becoming subsidy free, and it is therefore a challenge for wind farm developers to ensure good returns on investment. To investigate the future business cases for OWFs, two reference national scenarios are modelled namely NECP2030, based on the Dutch Integrated National Energy and Climate Plan (NECP) [2], and TRANSFORM where the Dutch society opts for aggressive structural changes towards a sustainable economy [12].

3.1. Modelling set up

Existing Energy System Models (ESM) are configured and set up to model the above scenarios. First, the OPERA ESM [13] is run under the NECP and TRANSFORM scenarios for 2030 and 2050 and its outputs (such as the hourly time series of demand for electricity and flexibility, the imports and exports) are used in the EYE model to obtain future prices in the Dutch electricity market. Additionally, the outputs of the future electricity prices from the EYE model are compared with results of COMPETES ESM [14] to have a robust understanding of trends of future prices in the electricity markets.

3.2. NECP2030

The installed capacity of offshore wind in 2030 will be 11.5 GW, and the Dutch electricity demand will be 137 TWh, not including flexible assets.

Table 1. Select inputs for NECP2030

Parameter	Value
Electricity demand	137 TWh
Flexible assets demand	30 TWh
PV capacity	20 GW
Onshore wind capacity	6.9 GW
Offshore wind capacity	11.5 GW
Natural gas capacity	17.8 GWh

From the price duration curve in Figure 1, the value of offshore wind in 2030 is 39.9 €/MWh. Consequently, the levelized cost of energy (LCoE) of OWFs must therefore be lower than this value to ensure a positive business case for wind farm owners.

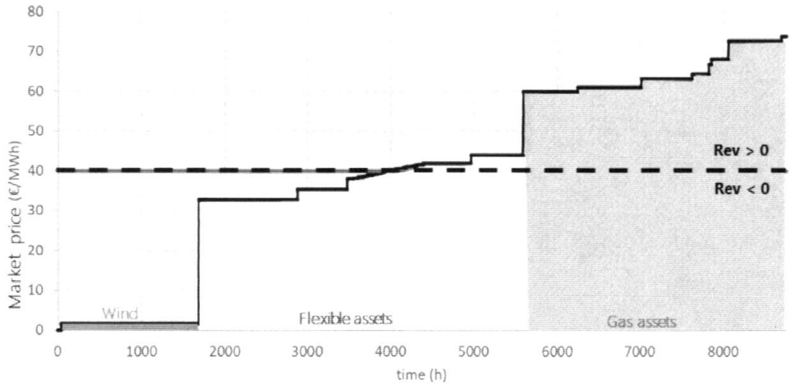

Figure 1. Market price duration curve for baseline NECP2030

Figure 1 shows that non-renewable assets will set the highest electricity prices in 2030. Also, flexible assets are the largest temporal price setters in the electricity market. The influence of flexible assets is further solidified by either removing them or doubling their capacity in two sensitivities. Table 2 shows that the value and utilisation of offshore wind decreases substantially without flexible assets, highlighting their importance.

Table 2. Influence of flexible assets on offshore wind value

Sensitivity	Value of offshore wind (€/MWh)	Offshore wind utilisation
Baseline flexible assets	39.9	100%
Remove flexible assets	23.5	55%
Double flexible assets	51.1	100%

3.3. TRANSFORM

This scenario envisions a society with radical behaviour and infrastructural changes towards a sustainable economy where electricity from solar and wind energy will become attractive for functions ranging from heating, mobility and industrial processes. Consequently, there is a substantial increase in both the electricity and flexible demand by 2050. The duration curves are characterized by higher prices, with an average of 51 €/MWh for 2030 and an extremely high price of 104 €/MWh in 2050. Figure 2 shows the market price duration curves of TRANSFORM 2030 and 2050 scenarios.

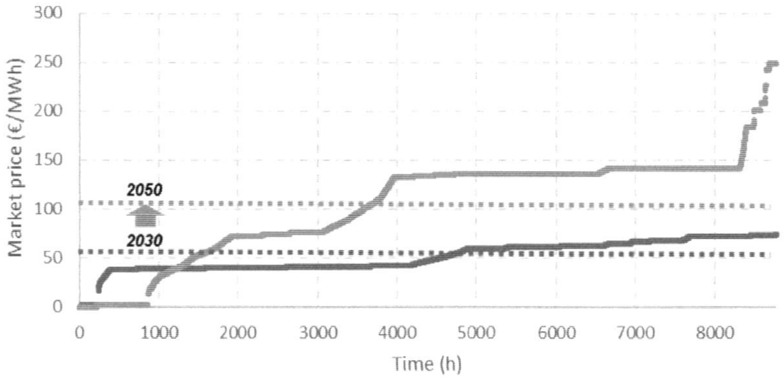

Figure 2. Market price duration curve for TRANSFORM 2030 and 2050.

4. Optimising the business case of OWFs

To improve upon the future value of offshore wind energy an example optimisation problem is defined. A reference 700 MW offshore wind farm is modelled with an electrolyser of the same capacity which under different scenarios uses profit maximization as the optimization objective. Table 3 lists inputs of the optimisation problem.

Table 3. Setup of offshore wind business case optimisation problem

Parameter	Value
Simulation length	3 months
Offshore wind asset capacity	700 MW
Offshore wind factor	48%
Maximum capacity electrolyzer	10 ton H2/h (1:1 wind-electrolyzer ratio)
Conversion factor electrolyzer	65 MWh/ton H2
Min. continuous operation electrolyzer	10% of capacity

4.1. Objective function

The objective function is to maximize the OWFs profit. The OWF has the choice of either selling wind energy to the spot market or to an electrolyser. It is optimized on an hourly basis and the profit is maximized over simulation length. The optimization problem is based on the model and equations developed in [15] , a study on the wind portfolio of the Spanish electricity market.

$$OWF\ profit = \sum_{h=1}^{h=8760} [(Revenues - Costs)_{day\ ahead\ h} + (Revenues - Costs)_{hydrogen\ h}] \tag{3}$$

In Figure 3, solid lines represent the flow of electricity (or chemicals) whereas dashed lines represent the flow of money. Green and red arrows indicate the revenues and the costs from the wind farm's perspective. W is the wind energy generation, DA is the day-ahead market, E_{da} is the energy bought from the electricity market to supply minimum requirements of energy to electrolyser when W is not enough and P is the price.

The revenues and costs for the offshore wind farm are:

$$Revenues\ wind_{da,\ h} = W_{da,h} * P_{da,h} \tag{4}$$

$$Costs\ wind_{\ day\ ahead,\ h} = W_{da,h} * LCOE_h \tag{5}$$

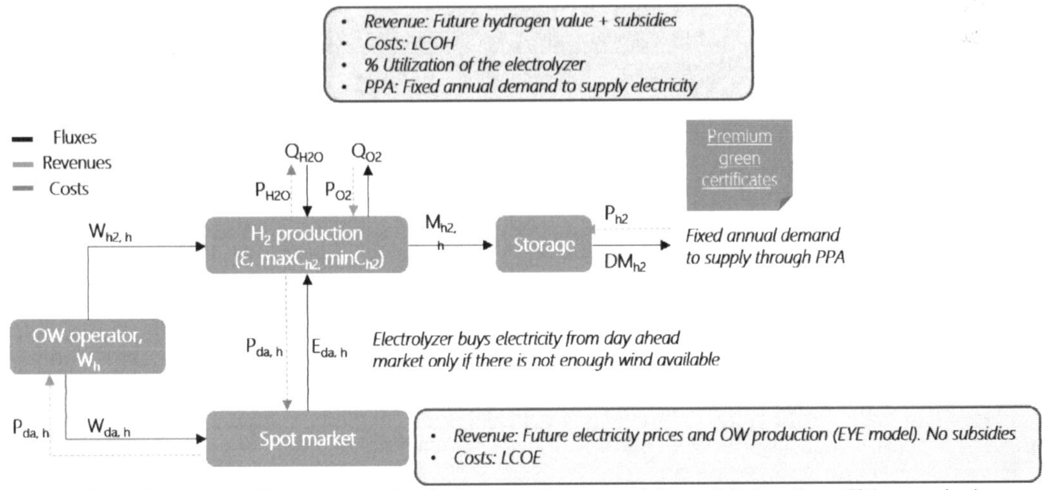

Figure 3. Flow diagram of the power-to-hydrogen business model combining the offshore wind energy generation and electrolysis.

4.2. Constraints

Profit for the OWF is obtained by either selling the wind energy in the electricity market or producing green hydrogen which is sold to third parties annually. A fixed annual supply of green hydrogen demand is modelled and the electrolyser is modelled at a minimum continuous load to avoid start-up times with a minimum utilization of 10%. Investment costs, capital expenditure and operational expenditure of the integrated model are analysed and taken into account as Levelized cost of Electricity (LCoE), when selling the wind energy in the electricity market and Levelized cost of Hydrogen (LCoH), when selling into the hydrogen market. In addition to the parameters in Table 3, for this case study, the cost of producing wind energy (LCoE) and cost of producing hydrogen (LCoH) are assumed to be 40 €/MWh and 3 €/kg_{H2} respectively. The constraints of the offshore wind energy generation, electricity market and green hydrogen production considered are:

$$W_{da,h} + W_{H2,h} = W \tag{6}$$

$$E_{da,h} + W_{H2,h} \leq Max\ capacity\ electrolyser\ (Max\ cap) \tag{7}$$

$$E_{da,h} + W_{H2,h} \geq Minimum\ electrolyser\ continuous\ load\ (Min\ load) \tag{8}$$

$$Annual\ DM_{H2} \geq Annual\ required\ green\ hydrogen\ demand\ to\ supply \tag{9}$$

4.3. Results and discussion

With the above optimisation setup, three cases are modelled as described in Table 4.

Table 4. Optimisation cases and results

Cases	Description	Offshore wind profits (M€)	Electrolyser capacity (tonH2/h)	Electrolyser utilisation
Case A	All OW energy to spot market	36.0	N/A	N/A
Case B	Optimise selling b/w spot market and electrolyser	41.4	10	18%
Case C	Downscale electrolyser capacity	39.6	3	35%

Selling all the energy in the spot market (Case A) results in a profit of 36.01 M€ for the OWF. When the OWF chooses to either sell to the spot market or to the electrolyzer (Case B), the profit increases to 41.4 M€. In Case B, the electrolyser has a capacity of 700 MW and is capable of producing 10 ton of hydrogen per hour. However the electrolyzer utilization is only 18% and to optimize this, its size is reduced by a factor of three. Table 4 shows an increase in electrolyzer utilization to 35%, but with reduced profits.

A further sensitivity analysis is performed, where the LCoE and LCoH are varied to see their influence on the electrolyser utilisation (Figure 4 left). The utilisation increases with low LCoH, this effect is justified as the more the electrolyser is used, the more the fixed investment costs are covered by the hydrogen production. Also, the LCoE seems to be indirectly proportional to the electrolyser utilization rate, meaning that when the LCoE is low the system prefers to sell the electricity to the spot market instead of feeding it to the electrolyser. Conversely, when the LCoE is high, the electrolyser is preferred, increasing its utilization. This is because the system looks at alternative ways to improve the value of offshore wind energy rather than selling it to the spot market.

Finally, the influence of this sensitivity on the offshore wind profit is seen in Figure 4 (right). The highest profit, 49.9 M€, is naturally obtained at the lowest operating costs of wind and electrolyser. The limit from which the profit starts to be positive is highlighted by the grey layer. Furthermore, from this figure it is clear that the variation of LCoE is more pronounced on the profits than the variation of LCoH.

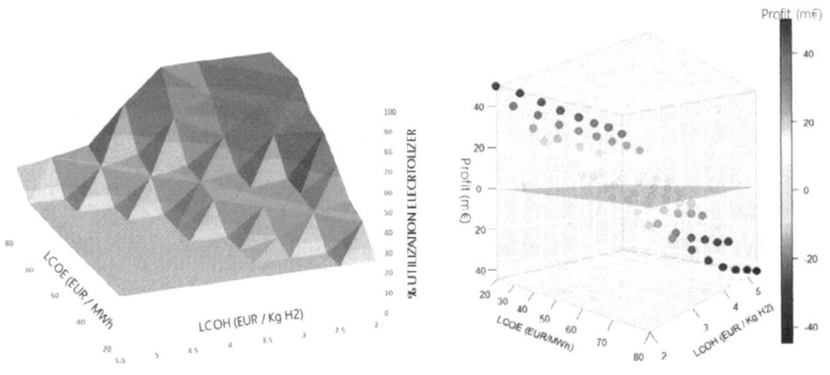

Figure 4. Electrolyzer utilization (%) depending on LCoE and LCoH (left). Optimization OWF profits results depending on LCoE and LCoH (right)

5. Conclusions and further work

In this work, the integration of high capacities of RES into the electricity market is explored. Results show that based on two national scenarios modelled in this study for 2030 and 2050, there will be a need for higher amounts of flexible assets such as batteries and electrolysers to support the business case for RES. Greater use of flexible assets will in turn increase electricity prices, with similar trends on future electricity prices observable in several European level studies of neighbouring countries.

Next, with the objective of maximising profits for a 700 MW wind farm, an optimisation problem is defined, where the wind farm has the choice to either sell electricity directly to spot market, or to an electrolyser. Results show that there is an increase in profit for the OWF when its energy is converted into hydrogen rather than being sold on the spot market.

A follow up for this study would be to model in greater detail the interaction between energy generation assets and flexible assets. In current energy system models, including the EYE model, it is possible to separately model various types of current energy generation and flexible assets. However, in the future energy system, hybrid or compound systems which can place bids on the electricity market as a single asset may be necessary. The business case for compound assets as well as their influence on setting the future prices of electricity will be explored in subsequent studies.

References

[1] European Commission 2021 *'Fit for 55': delivering the EU's 2030 Climate Target on the way to climate neutrality* External Brussels: European Commission Report No: 52021DC0550

[2] Abels-van Overveld M, Bleeker A and Boot P 2019 *Climate and Energy Outlook 2019* The Hague: PBL Netherlands Environmental Assessment Agency Report No: 3825

[3] Wiggelinkhuizen E and Eecen P 2018 *Flexibility options for market-exposed offshore wind farms in the North Sea*

[4] Stappel M, Gerlach AK, Scholz A and Pape C 2015 *The European Power System in 2030: Flexibility Challenges and Integration Benefits* Kassel: Fraunhofer-Institute for Wind Energy and Energy System Technology (IWES) Report No: 067/02-A-2015/EN

[5] Renaud A, Khallouf P and Andrey C 2019 *Determination of a target electricity interconnection capacity between France and the United Kingdom* Paris: Artelys Solutions and Optimisation

[6] Graichen P 2018 *Energiewende 2030: The Big Picture* Berlin: Agora Energiewende

[7] Ingwersen J and Schmitt L 2020 *Ten-Year Network Development Plan (TYNDP)* Entso-E

[8] Bossmann T, Fournié L and Verrier GP 2018 *METIS Studies* Brussels: European Union Renewables, Research and Innovation, Energy Efficiency

[9] Barges R, Bossavy A, Chammas M, Fournié L, Khallouf P and Texier B 2016 *METIS Technical Note T2* Brussels: European Commission Renewables, Research and Innovation, Energy Efficiency

[10] Matthijsen J, Dammers and Elzenga H 2018 *The future of the North Sea: the North Sea in 2030 and 2050 (scenario study)* The Hague: Report No: 3193

[11] Bot ETG 2015 *FarmFlow validation against full scale wind farms* Petten, the Netherlands:

[12] Scheepers M, Gamboa Palacios S, Jegu E, Nogueira De Oliveira LP, Rutten L, van Stralen J, et al. 2020 *Towards a sustainable energy system for the Netherlands in 2050* Public The Hague: Report No: TNO 2020 P10338

[13] de Joode J, Dalla Longa F, Smekens K and Daniels B March 2016 *Integrating energy systems at the regional level: A model based assessment of the Dutch energy system*

[14] Sijm S, Gockel G, van Hout M, Smekens K, van der Welle A, van Westering W, et al. November 2017 *The supply of flexibility for the power system in the Netherlands, 2015-2050*

[15] González-Aparicio I, Pérez-Fortes M, Zucker A and Tzimas E 2016 Opportunities of Integrating CO2 Utilization with RES-E: a Powerto-Methanol Business Model with Wind Power Generation *Elsevier* **114** 6905 – 6918

Offshore wind farm wake modelling using deep feed forward neural networks for active yaw control and layout optimisation

S Anagnostopoulos [1] and MD Piggott [2]

emails: sa1619@imperial.ac.uk, m.d.piggott@imperial.ac.uk

Abstract. Offshore wind farm modelling has been an area of rapidly increasing interest over the last two decades, with numerous analytical as well as computational-based approaches developed, in an attempt to produce designs that improve wind farm efficiency in power production. This work presents a Machine Learning (ML) framework for the rapid modelling of wind farm flow fields, using a Deep Neural Network (DNN) neural network architecture, trained here on approximate turbine wake fields, calculated on the state-of-the-art wind farm modelling software FLORIS. The constructed neural model is capable of accurately reproducing single wake deficits at hub-level for a 5MW wind turbine under yaw and a wide range of inlet hub speed and turbulence intensity conditions, at least an order of magnitude faster than the analytical wake-based solution method, yielding results with 1.5% mean absolute error. A superposition algorithm is also developed to construct flow fields over the whole wind farm domain by superimposing individual wakes. A promising advantage of the present approach is that its performance and accuracy are expected to increase even further when trained on high-fidelity CFD or real-world data through transfer learning, while its computational cost remains low.

1. Introduction

Wind power has maintained a significant share of the total worldwide electrical energy production of the 21st century (4.8% by the end of 2018 [1]) and is expected to grow up to 18% within the next 3 decades [2]. Wind energy production has increased over the past few years due to extensive efforts towards the modelling of turbine wakes and wind farm configurations that include a variety of analytical, experimental as well as high-fidelity numerical approaches.

Analytical wake modelling involves analytical or semi-analytical techniques used to perturb background wake flow information, providing an estimate for the flow field in the presence of turbines. Analytical modelling has been present from the earliest stages of modern wind turbines and is still playing an important role in producing low-cost wake predictions of the wake deficit profile. Some of the most commonly used models such as Larsen [3], Jensen [4], Curl [5] and Gaussian [6] are usually incorporated in packages available in the industry (FLORIS, WasP, WindPro etc) and depending on the application, they can significantly reduce the complexity of parametric studies or turbine array set-ups. However, since they usually involve the use of highly simplified physical assumptions and are mainly focusing on averaged velocity profiles and not transient turbulent wakes, they do not constitute a method that produces results of high-accuracy [7], [8]. Furthermore, the analytical models rely on empirical constants which require fine-tuning through computationally-expensive CFD simulations. Although these models are not able to capture a detailed representation of the turbine velocity deficit, the trade-

[1] Corresponding author, Mechanical Engineering, EPFL, Lausanne, Switzerland
[2] Department of Earth Science & Engineering, Imperial College London, UK
Github: https://github.com/soanagno/wakeNet

off between accuracy and low-computational time is very often made in support to high-fidelity CFD studies in optimisation problems.

During the last two decades, the exponential advancements in CPU as well as GPU architectures have led to a significant progress towards wake modelling using computational methods [9], [10]. Coupled numerical models such as large-eddy simulation (LES) and Reynolds-averaged Navier-Stokes (RANS) are now capable of describing steady as well as transient wake flows very accurately and provide a realistic illustration of wind-turbine interaction physics [11], [12]. Nevertheless, accurately representing the wind flow profile, especially through the turbine blades which are moving at very high rotational speeds, requires very fine mesh settings in order to capture the boundary layer properties. Even then, some of the most widely used models for turbulence like k-ε, tend to overestimate the turbulence viscosity [13] leading to discrepancies between the numerical results and experimental measurements [14]. Thus, CFD is rarely used as a standalone method, even less when dealing with array optimisation problems (e.g. using Adjoint methods) which require multiple high-computational-cost runs.

Through the recent increase of computational power, the application of Artificial/Deep Neural Networks (ANN, DNN) as well as Convolutional Neural Networks (CNN), has been successfully tested across a wide variety of scientific fields, including fluid mechanics [15], [16], [17]. On wind turbines, there have been some recent efforts towards modelling of source terms with an ANN [18] and correlating Reynolds stress anisotropy with strain [19]. An ANN was constructed in order to assess the performance of wind turbines using the power vs torque curves [20], [21]. Very recently, a simple ANN architecture was trained on a large high-fidelity dataset and correlated two inputs (inlet wind speed and turbulence intensity) to produce 3D wake profiles of wind turbines in a single row [22]. Computationally efficient surrogate modelling has also been used to estimate long-term wake-induced loading effects and turbine lifetime, while leading to AEP improvements of up to 5% in various wind farm layout optimisation scenarios [23], [24], [25]. So far, in the limited available research, neural networks appear to be very efficient in the challenging task of building relationships between inflow conditions, rotor specifications and fluid properties. Traditional methods face difficulties in producing fast results that at the same time are accurate enough to support parametric studies. A well-constructed and well-trained neural network could be deployed and provide reliable results within seconds in order to predict flow properties of a wind farm that would otherwise require orders of magnitude higher computational times. At the moment, the use of powerful machine learning and regression tools shows promising results and could pave the way to even more sophisticated optimisation approaches in the future.

The present study aims at constructing a neural model, capable of handling wind properties and wind turbine yaw settings, to demonstrate that active yaw control and layout optimisation using ML tools for turbine wake modelling is feasible, at considerable computational time gains. Using the presented neural network framework could enable further accuracy and computational cost improvements in wind farm optimisations that could easily be implemented with transfer learning (using gained ML knowledge from one problem to a similar one), with more advanced wake datasets, produced by iterative analytical models and high-fidelity numerical simulations (CFD) for further training the DNN.

2. Software Description & Methodology

A Deep Feed Forward (DFF) Neural Network is built in order to reproduce the velocity domain of a single wind turbine wake, when given up to a triplet of inlet conditions: upstream velocity at the hub (inlet speed), turbulence intensity and yaw angle. The network is trained on wake profiles produced by FLORIS (v2.1.1) using the Gaussian analytical wake model, where the 3D velocity deficit behind each turbine is derived from the following simplified form of Navier-Stokes:

$$\frac{u(x,y,z)}{U_\infty} = 1 - Ce^{(y-\delta)^2/2\sigma_y^2}e^{-(z-z_h)^2/2\sigma_z^2} \tag{1}$$

$$C = 1 - \left(1 - \frac{(\sigma_{y0}\sigma_{z0})M_0}{\sigma_y \sigma_z}\right)^{1/2} \tag{2}$$

$$M_0 = C_0(2 - C_0) \tag{3}$$

$$C_0 = 1 - \sqrt{1 - C_T} \tag{4}$$

where C is the velocity deficit at the centre of the wake, δ is the wake deflection, z_h is the hub height, C_T is the thrust coefficient and σ_y, σ_z the widths of the wake in the y and z direction, respectively. The "0" subscript denotes the values at the start of the far wake and the widths σ_y, σ_z are given by:

$$\frac{\sigma_y}{D} = k_y \frac{x - x_0}{D} + \frac{\sigma_{y0}}{D} \tag{5}$$

$$\frac{\sigma_z}{D} = k_z \frac{x - x_0}{D} + \frac{\sigma_{z0}}{D} \tag{6}$$

where

$$\frac{\sigma_{y0}}{D} = k_y \frac{x - x_0}{D} + \frac{\sigma_{z0}}{D} \cos(\gamma) \tag{7}$$

$$\frac{\sigma_{z0}}{D} = \frac{1}{2}\left(\frac{u_R}{u_\infty + u_0}\right)^{1/2} \tag{8}$$

where D is the rotor diameter, γ is the yaw angle and k_y, k_z define the wake expansion in the lateral and vertical directions, respectively.

2.1 Synthetic dataset method

The synthetic dataset was comprised of 4000 wake images (Fig. 1), with a batch size of 400 wakes, which was found to be sufficient in producing results of satisfactory accuracy (Fig. 4) for the purposes of the present study, and was applied on a 5 MW wind turbine of 126m in diameter. A batch size of 10% of the complete dataset lies within the commonly adopted range in the literature [26] and was proven to be effective in reducing overfitting of the model to specific wake inputs, while also reducing the training computational cost.

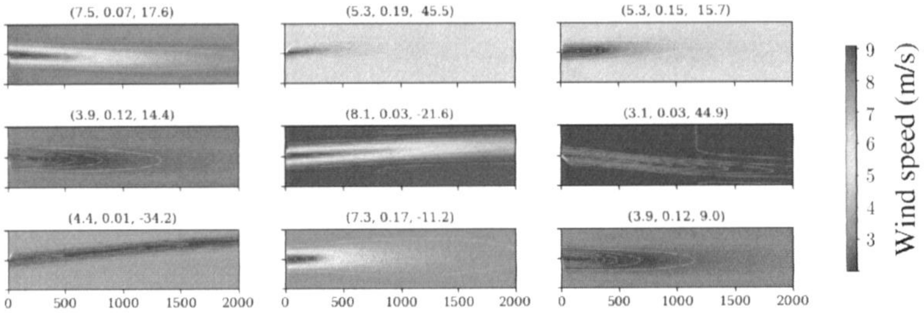

Figure 1. Indicative wake dataset sample where each title denotes a wind speed (ws), turbulence intensity (ti), yaw angle (yw) triplet.

The ranges of the inlet speed (ws), turbulence intensity (ti) and yaw angle (yw) are [3, 12] m/s, [0.01, 0.2] and [-50, 50] degrees, respectively, while their values are produced randomly following a uniform distribution. The range of the inlet speeds is selected based on the power curve of the specific wind turbine (shown in Fig. 2), where the lower operation limit is 3 m/s and the upper limit is 20 m/s. However, right after the point where the Cp value becomes constant (at 12 m/s), there is usually little potential power gain from yawing the hub [27], thus the selected upper limit for the training was set at 12 m/s, which significantly reduced the size of the required synthesised dataset. The range of the turbulence intensity is selected based on commonly reported measurements [28]. It is worth mentioning that the range, as well as the distribution of the dataset, could easily be adjusted to follow a non-uniform distribution based on available weather data, in order to further improve the capabilities of the DNN on site-specific weather conditions.

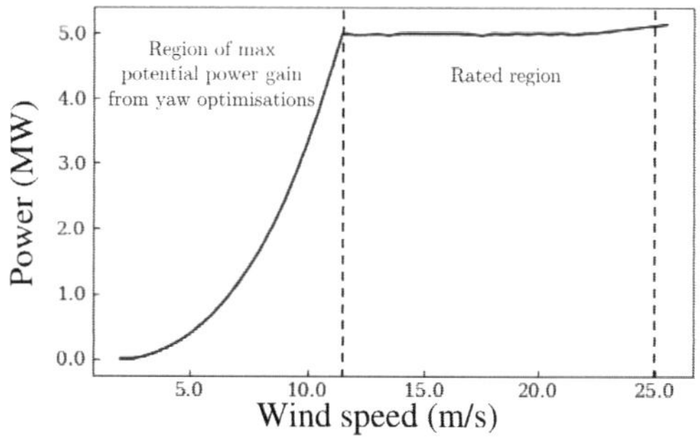

Figure 2. Power curve of the 5MW wind turbine of the study. The rated region of the turbine is the "plateau" of potential power output and begins after about 12 m/s.

2.2 DNN architecture

The architecture of the DNN is shown in Figure 3, which consists of at least two hidden layers of 100 neurons each, that output an m by n grid of points that represent the downstream velocity domain of the wake. It has been shown that re-normalising the training batch after each hidden layer of a DNN significantly increases the performance of the trained network [29]. Thus, two batch normalisations have been applied after the first two layers, which as expected, increased the accuracy of the resulting wake. Several activation functions were tested for their performance (sigmoid, tanh, tansig, purelin), as well as various normalisation methods (min-max, mean/std, -1/1) for the input parameters. The selected activation functions which performed best and thus is selected for this study are the tanh for the first two layers and a linear activation function for the output along with mean/std normalisation.

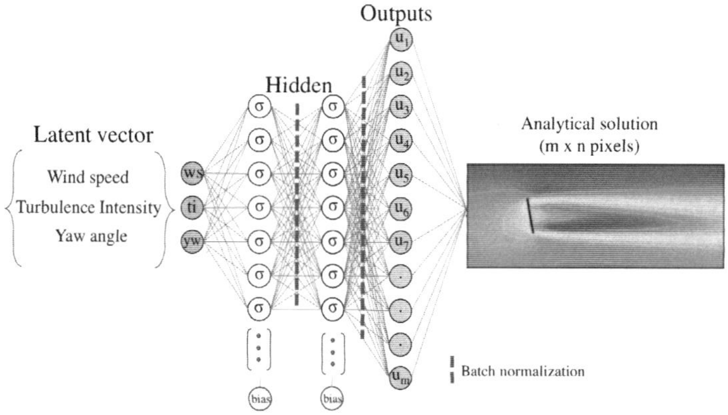

Figure 3. DNN architecture. The dots represent the optional functionality of independent parallel sub-networks, each of which is trained on a partition of the domain.

The default resolution of FLORIS is a 200 x 200 grid which is also the default adopted resolution for the training of the DNN. However, for higher resolution demands, the user can specify the desired resolution before training the model which will also produce an analytical wake dataset of that same resolution. To tackle arising memory problems for models trained on higher resolution images, a parallel sub-network functionality has been implemented (Fig. 3). These sub-networks are deployed in parallel instead of the main network and are trained independently on a portion (piece) of the domain, specified by the user. Furthermore, this functionality allows the user to specify if each independent "piece" of the domain will be row, column or block-wise.

2.3 Hyper-parameters & training

The model hyperparameters (Table 1) were calibrated through iterative process by evaluating the metrics of the loss and absolute accuracy of the validation set (Figs 4b, 4d). A common practice when using large datasets is to create a validation test that constitutes 10-20% of the initially created dataset [30]. The training dataset, is the remaining proportion of the synthesised FLORIS images (90%), and is compared with the DNN's output images by applying the Root Mean Square Error (RMSE) between the wake velocities at each cell of the computational grid, which is given by:

$$RMSE_{fo} = \sqrt{\frac{1}{N}\sum_{i=1}^{N}(z_{fi} - z_{oi})^2} \quad (9)$$

where z_{fi}, z_{oi} are the DNN forecasts and analytical observed values, respectively and N is the sample size. Note that for the final training, the whole dataset (4000 wake images) is used as the training dataset.

Several optimisers were tested for their performance including Stochatic Gradient Descent (SGD), RMSprop, Adam and Resilient Back Propagation (Rprop) [31]. The latter, qualified as the fastest converging optimizer for this study and was one of the most significant improvements on early vanilla versions of the DNN's development, resulting in a ~97% accuracy on both training and validation datasets, as shown in Figure 4.

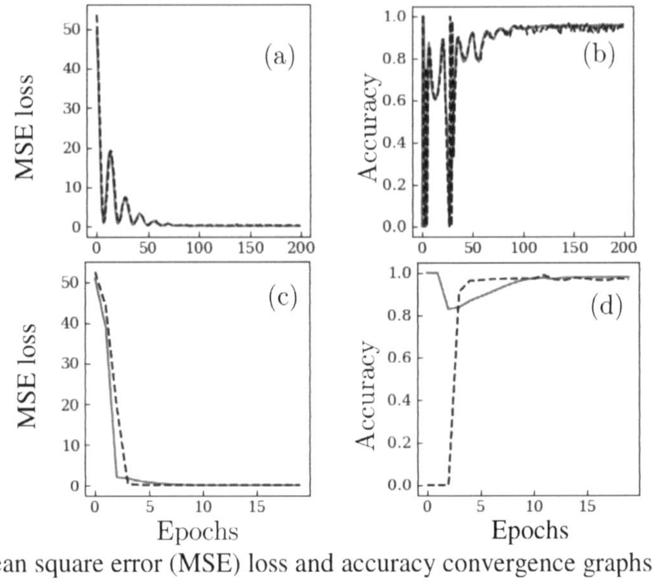

Figure 4. Root mean square error (MSE) loss and accuracy convergence graphs of Stochastic Gradient Descent (SGD) (a, b) and Resilient Back Propagation (Rprop) (c, d), respectively.

Table 1. Model hyper-parameters, MSE loss, accuracy and timings for SGD and Rprop optimisers.

	Learning rate	Momentum	Validation Loss	Training Loss	Validation Accuracy	Training Accuracy	Time (m)
SGD	0.1	0.99	0.06	0.2	0.95	0.94	50
Rprop	0.01	-	0.035	0.05	0.98	0.97	15

2.4 Wake superposition

Most analytical wake models include an independent approach for the superposition of wakes in order to form an array of multiple wind turbines, as well as the interactions between them [32], [33].

For this study, a superposition algorithm has been developed based on the SOS model, which is deployed for the combination of multiple individual wakes produced by the DNN. The superposition is modified with an approximate method of calculating a uniform velocity at the hub of each turbine. The final domain represents a collage of the individual wakes that comprise the examined offshore wind farm.

One of the advantages of the presented neural model is that it is trained on 2D slices of a 3D wake domain of FLORIS, which also implicitly consider the vertical velocity boundary layer, as affected by the sea surface. This significantly contributes in faster computation times, while including information about the velocity boundary layer of the site. However, since only the velocities along the line of the hub are known, an approximate formula has to be applied in order to correct the average velocity taken at the hub of each turbine. As shown in Figure 5, the turbine hub can be divided in radial rings of equal thickness dr, which represents the length of one pixel of the computational grid.

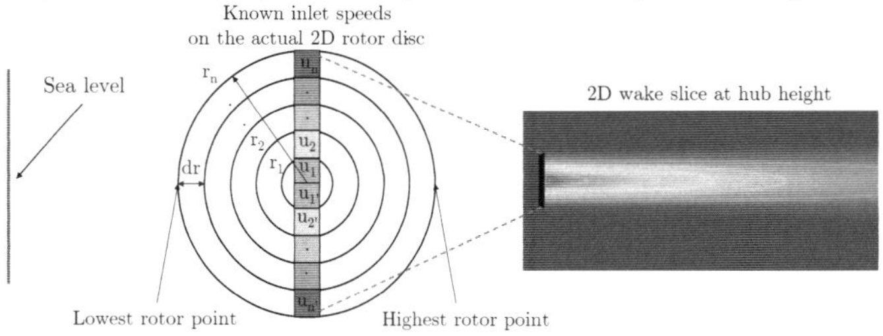

Figure 5. 2D representation of the 1D hub as calculated by the DNN.

Assuming that the rotor lies within a linear region of the boundary layer, by integrating over the surface of the hub, the velocity on the 2D disc is given by:

$$u_{hub} = \int_{r_1}^{r_{n/2}} u_{mean} \cdot 2\pi r \cdot dr \quad (10)$$

where u_{mean} is the mean speed on each hub ring and the power of each turbine is given by:

$$P = \frac{1}{2} Cp \cdot \rho \cdot \cos\theta \cdot {u_{power}}^3 \cdot A \quad (11)$$

where Cp is the turbine's power coefficient and ρ is the fluid density and θ is the yaw angle of the wind turbine. Therefore, for a farm of n turbines, the total generated power is given by:

$$P_{tot} = \sum_{i=1}^{n} P_i \quad (12)$$

2.5 Yaw and layout optimisations

Recent studies [27], [34] have shown that exploiting the active yaw control of wind farms using analytical models like FLORIS, can increase the annual energy production (AEP) of the wind plant and also reduce the thrust loads on the turbines. Fast active yaw optimisation using machine learning could be applied at a minimal additional cost, boosting the AEP even further since the reaction time of the turbine could be significantly decreased.

On the other hand, positioning the wind turbines such that the expected power production is maximised (Wind Farm Layout Optimization Problem, WFLOP [35]) is another complex problem that the existing available literature has very recently started attempting to solve. Analytical, as well as high-fidelity CFD simulations, have been previously used along with Genetic Algorithms (GAs) [36] and Adjoint methods [37], [38], [39] and although they represent a good starting point towards the WFLOP problem they usually constitute costly computational approaches. Moreover, they lack multi-objective optimisation aspects like the effect of turbulence intensity on the plant or other area-specific construction limitations [35].

On these grounds, two distinct optimisation modules are also developed in this work in order to assess the capability of the neural wake model in wind farm active yaw and array turbine placement optimisation problems. The optimisation functions are made from scratch using SciPy's SLSQP optimiser and attempt to optimise the total power output of the plant (Eq. 12) by handling the outputs produced by the DNN. The results of the optimisers are then compared with the corresponding optimised results of FLORIS, for both yaw and layout optimisation scenarios. The reliability and the accuracy of the modules is finally assessed for the whole operation range of the DNN.

3. Results

To demonstrate the capabilities of the trained neural model, several indicative cases are examined, namely two cases (single and multiple wakes) for the assessment of the mean absolute error (%) between FLORIS and the DNN, two cases of multiple wakes to assess the correctness in the implementation of the SOS superposition model and three optimisation cases (A, B, C), the first two for the yaw and the latter for the optimisation of the farm layout. All power gain comparisons were performed using FLORIS, with the proposed optimal settings of either FLORIS or the DNN.

3.1 Single / multiple wake plots with errors and y-transects

Figures 6a-c and 7a-c show a single wake and a multiple wake scenario, respectively. The absolute relative error (%) between the analytical and neural results is shown in Figures 7c and 8c, and is given by:

$$|error_{\%}| = \frac{1}{n}\sum_{i=1}^{n}\left|\frac{u_{fi}-u_{oi}}{u_{fmax}}\right| \cdot 100 \quad (13)$$

where u_{fi}, u_{oi} are the velocities calculated at each pixel i by FLORIS and by the DNN, respectively, u_{fmax} is the maximum velocity of the FLORIS domain and n is the total number of pixels (200x200 by default).

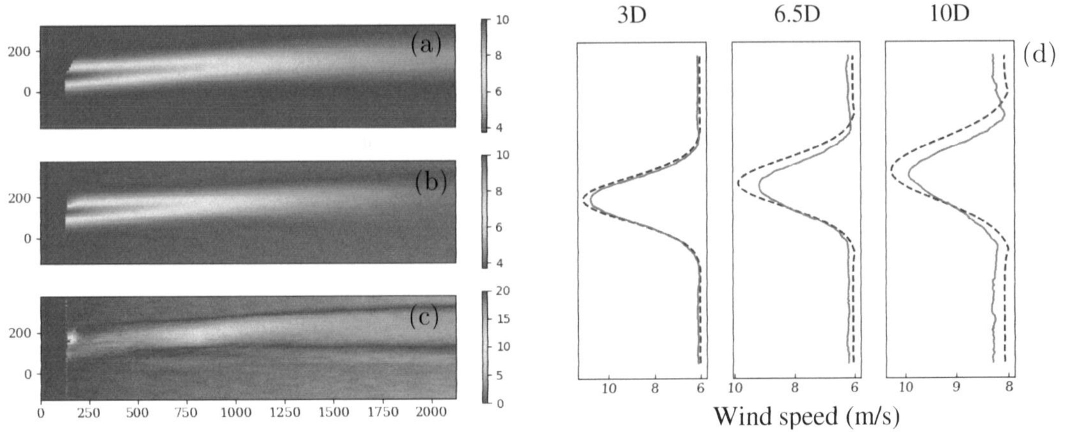

Figure 6. Single wake produced by FLORIS (a), the DNN (b). Relative absolute error (%) between the two models (c) and y-transects at 3, 6.5 and 10D downstream (d). The inlet conditions are ws: 11 m/s, ti: 0.05, yw: 30°. DNN: red line, FLORIS: blue dashed line.

For the single wake case (Fig. 6), the resulting DNN data appears to reproduce well the analytical wake deficit, with the highest error being at 10%, while the DNN velocity profiles at three y-transects along vertical cross-sections of the physical domain agree with the analytical ones in terms of the yawing angle, showing some minor deviation the further they are from the hub (Fig. 6d). The mean absolute error is calculated for a 400 different wakes of random inlet conditions as produced by the neural model and FLORIS, which was found to be at 1.5%.

The multiple wake case (Fig. 7) was selected in order to test the performance of the DNN, this time on a dense wind farm configuration with high turbine yaw settings. As expected, the absolute error increases (namely up to 20%) in certain regions of the wake velocity domain (Fig. 7c), since the superposition method is now also affecting the results. However, the mean absolute error is less than 10%, while the y-transects in Figure 6d agree well between the two models with only few exceptions, mainly in regions of multiple wake superposition.

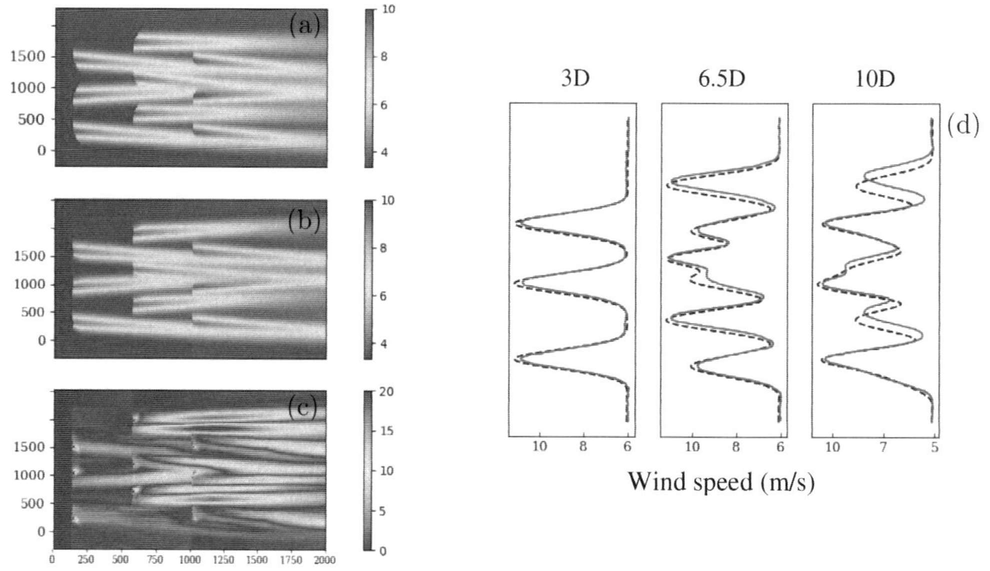

Figure 7. Multiple interacting wakes produced by FLORIS (a), the DNN (b). Relative absolute error (%) between the two models (c) and y-transects at 3, 6.5 and 10D downstream (d).

3.2 Scalability for wind farm with absolute error

Regarding the computational cost, in Figure 8 the increase of computational cost with the number of turbines appears to be of 2nd order for Floris, and 1st order for the DNN, resulting in 3 orders of magnitude gain for an array of 20 turbines. Note that the computational time for each number of turbines is obtained by taking the average time of 5 iterations of the same superposition scenario.

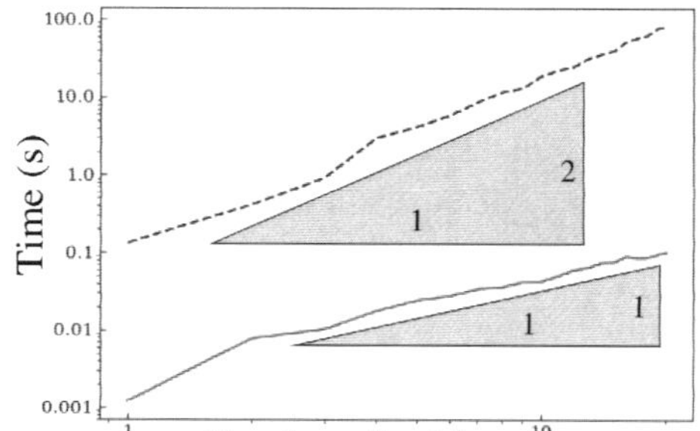

Figure 8. Logarithmic plots of computational time scaling vs number of superimposed turbines for FLORIS (blue dashed), DNN (red). The triangle sides indicate the orders of magnitude.

3.3 Scalability for wind farm with absolute error

Yaw optimisation has recently become a key topic, especially in wind farm configurations where the total power output of the plant can be significantly boosted based on the yaw settings of the turbines

[27]. Furthermore, active yaw optimisation is heavily dependent on fast predictions based on live weather forecasting [40], [41] as to take advantage of as much of the incoming wind energy as possible. This study attempts to demonstrate that even when compared to a relatively cheap wake model as the Gaussian, a DNN model can produce very good yaw setting predictions with significant computational cost gains, given that weather conditions are known. Two indicative example cases are discussed in this section (A and B), involving a 6- and a 15-turbine wind farm configuration, respectively.

In general, compared to a greedy yaw setting where the turbines are facing the wind direction (zero yaw in the present case), deflecting the wakes of the front turbines can usually be regarded as a good yaw setting strategy for a set of turbines that cooperatively increase the overall performance of the farm [34]. In Figure 9a, FLORIS achieves a 25.8% power gain by setting a 29, 18, 0° of yaw on the three turbine pairs from front to back of the stream, respectively, after 15s. A very similar approach is adopted by the optimisation produced by the DNN, where a 22.8% power gain is achieved with a 26, 20, 0° corresponding setting. The optimisation required less than half the time (7s). Note that for all yaw optimisations, the initial yaw angles are at 0 degrees.

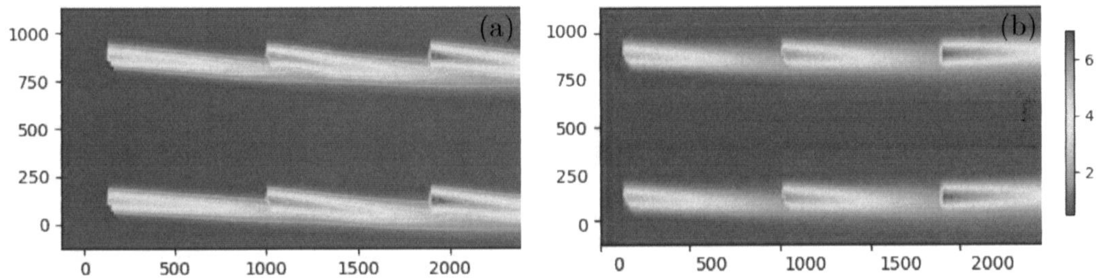

Figure 9. Case A: Optimised yaw result for a 2x3 wind farm array deduced by FLORIS yaw optimisation module (a) and the DNN (b).

Table 2. Optimised results produced by FLORIS and the DNN for Case A. Yaws correspond to turbines from left to right and bottom to top.

	Yaws (deg)	Initial power (MW)	Optimal power (MW)	Power gain (%)	Comp. time (s)
FLORIS	[29, 29/ 18, 18/ 0, 0]	3.77	4.75	25.8	15
DNN	[26, 26/ 20, 20/ 0, 0]	3.77	4.63	22.8	7

Case B examines the yaw optimisation of a higher density 15-turbine wind farm, where by using a similar yaw cooperative strategy the power gain achieved by the DNN is again very close to that of FLORIS (28.7% and 31.6%, respectively), requiring almost half the computational time (60s vs 110s).

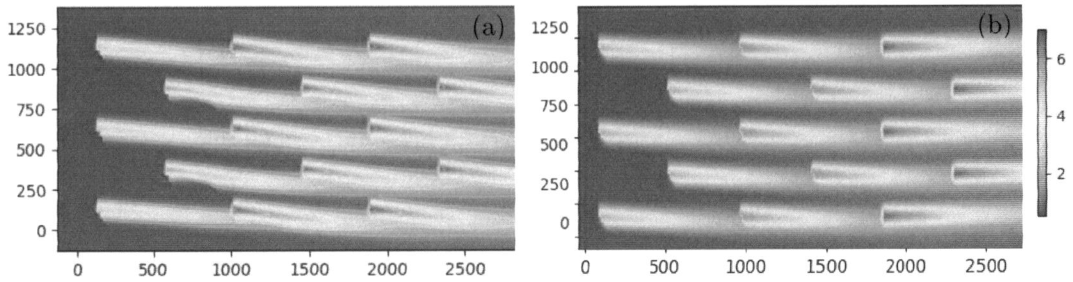

Figure 10. Case B: Optimised yaw result for a 15x wind farm array deduced by FLORIS yaw optimisation module (a) and the DNN (b).

Table 3. Optimised results produced by FLORIS and the DNN for Case B. Yaws correspond to turbines from left to right and bottom to top.

	Yaws (deg, from left to right rows)	Initial power (MW)	Optimal power (MW)	Power gain (%)	Comp. time (s)
FLORIS	[30, 30, 30/ 23, 23/ 8, 8, 8/ 7.5, 7.5/ 0, 0, 0/ 0, 0]	9.41	12.39	31.6	110
DNN	[25, 24, 24.5/ 24.5, 24/ 19, 18, 18/ 18, 18/ 0, 1, 1/ 0, 0]	9.41	12.12	28.7	60

3.4 Yaw optimisation - Farm power heatmaps

For the above cases A and B, the corresponding total farm power heatmaps are presented in Figures 11 and 12, in order to assess the capability of the neural network in obtaining the optimal power gain. As evident from these heatmaps, the region with the highest potential power gain is around the bottom right corner, which is defined by low turbulence intensity and high inlet speeds. Although cases where high wind speeds are not usually accompanied by low turbulence intensity [27], they are examined for the purpose of creating a heatmap across the whole range of inlet conditions. For Case A, the DNN is capable of producing at least 75% of the total power gain of FLORIS (~80% for speeds below 8m/s), as shown in Figure 11b within 1/10 of the computational time (60s vs 5s). Similarly, in Case B, the DNN achieves a large proportion of the total power gain of FLORIS, requiring less than 1/10 of the computational time (180s vs 16s).

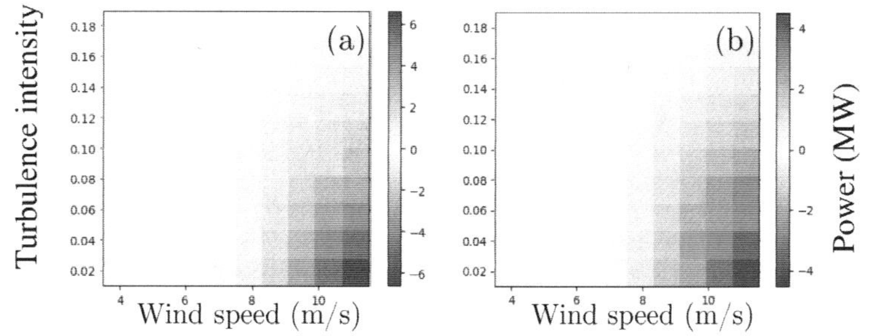

Figure 11. Case A: Optimised yaw heatmaps for FLORIS (a) and the DNN (b) with ti and ws ranges [0.01-0.19] and [3.5-11.5], respectively. Average FLORIS time: 60s, average DNN time: 5s.

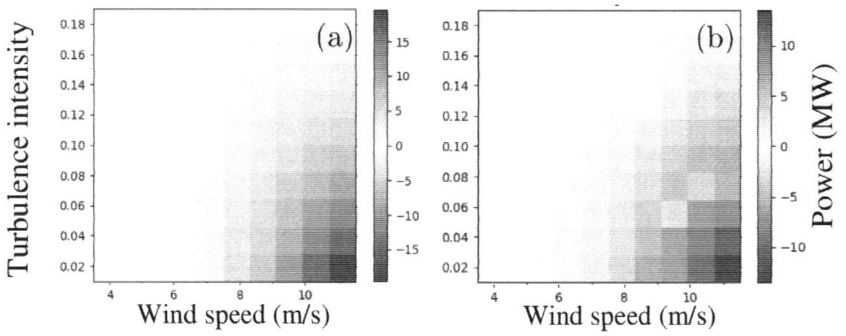

Figure 12. Case B: Optimised yaw heatmaps for FLORIS (a) and the DNN (b) with ti and ws ranges [0.01-0.19] and [3.5-11.5], respectively. Average FLORIS time: 180s, average DNN time: 16s.

3.5 Layout optimisation

An indicative test case for layout optimisation of the initial configuration (Fig. 13a) is presented below, where the wind direction is assumed to be constant at 7 m/s and the boundary constraints of the domain are 20D X 10D. A minimum lateral distance between each individual turbine of 2D has also been implemented as a constraint in the code. As shown in Figure 13b, the optimum configuration obtained with the DNN model achieves a 78.93% power gain (or AEP gain), whereas FLORIS achieves 79.41%

(Fig. 13a). The computational time required by the DNN to provide a forward solution is an order of magnitude less than that of FLORIS (5.5 sec vs 56 sec). Any difference between the optimal layouts in Figure 13 can be attributed to the fact that the solution is not unique, i.e. that multiple designs have similar power outputs and the optimisation algorithm may converge to either of these.

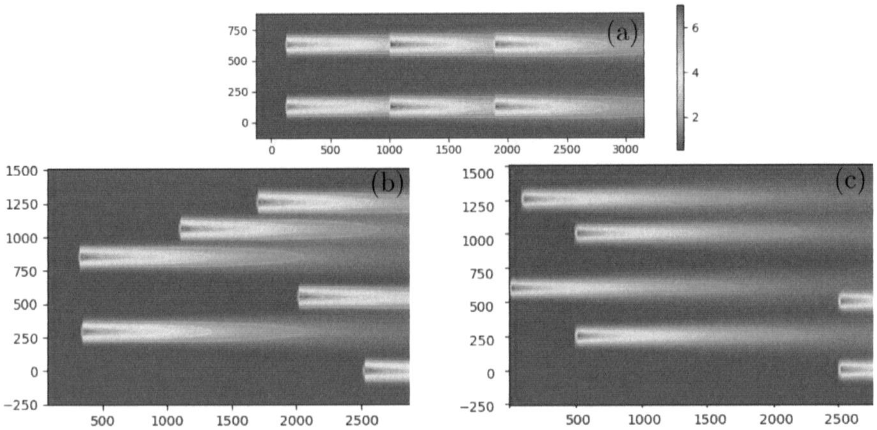

Figure 13. Case C: Initial 2x3 layout configuration (a), optimised layout result produced by FLORIS Layout Optimisation module (b) and the DNN model (c).

Table 4. Optimised results produced by FLORIS and the DNN for Case C.

	Initial power (MW)	Optimal power (MW)	Power gain (%)	Comp. time (s)
FLORIS	3.78	6.78	79.41	56
DNN	3.78	6.76	78.93	5.5

4. Conclusions

In this work, the capabilities of machine learning on the modelling of offshore wind farm wakes are investigated. The main stages and corresponding accomplishments of this study are the following:

a. Development of a DNN for the accurate representation of 2D offshore turbine wakes under yaw, which produces the velocity fields of single wakes with a mean absolute error of 1.5%. The 2D velocity deficits at hub-level produced by the DNN at the hub level also include information of sea-level effect on the wind boundary layer downstream wind.
b. Implementation of a superposition method, which combines the SOS model along with an analytical approximation that renders the 2D wake deficit produced by the DNN sufficient in predicting the power output of a given wind farm. The scaling of the computational time for the superposition of up to 20 turbines was found to be up to three orders of magnitude faster.
c. Demonstration of the optimisation capabilities of the neural model against those of FLORIS. For the yaw optimisation, the optimal yaw settings produced by the DNN provides at least 75% of the farm power output of the corresponding FLORIS optimisation module, on average 10 times faster, which can enable power gains due to the faster reaction time of the yaw control.
d. Development of an open-source machine learning wake model with various set-up options, aiming to provide flexibility to the user, depending on the needs of application. Using the default model hyper-parameters and settings, a reliable neural wake model can be trained on a personal computer, requiring less than 15 minutes of training time. Transfer learning could be applied in future versions, using high-fidelity CFD runs to further increase the model's capabilities, while maintaining the same low order for the evaluation time of the network.

5. References

[1] Council GWE, Annual Wind report (2019)
[2] Council GWE, Global wind report annual market update (2013), GWEC, Brussels
[3] G. C. Larsen (2009), A simple stationary semi-analytical wake model, Riso National Laboratory for Sustainable Energy, Technical University of Denmark
[4] N. Jensen (1983), A note on wind generator interaction
[5] L.A. Martinez-Tossas, J. Annoni, P.A. Fleming, M.J. Churchfield (2019), The aerodynamics of the curled wake: a simplified model in view of flow control, *Wind Energ. Sci.,* 4, pp. 127-138
[6] M. Bastankhah, F. Porté-Agel (2014), A new analytical model for wind-turbine wakes, *Renew Energy*, 70, pp. 116-123
[7] N. Sedaghatizadeh, M. Arjomandi, R. Kelso, B. Cazzolato, M.H. Ghayesh (2018), Modelling of wind turbine wake using large eddy simulation, *Renew Energy*, 115, pp. 1166-1176
[8] A. Niayifar, F. Porté-Agel (2016), Analytical modeling of wind farms: A new approach for power prediction, *Energies*, 9, p. 741
[9] Y.-T. Wu, F. Porté-Agel (2012), Atmospheric turbulence effects on wind-turbine wakes: An LES study, *Energies*, 5, pp. 5340-5362
[10] M.P. van der Laan, N.N. Sørensen, P.E. Réthoré, J. Mann, M.C. Kelly, N. Troldborg, *et al.* (2015), An improved k-ε model applied to a wind turbine wake in atmospheric turbulence, *Wind Energy*, 18, pp. 889-907
[11] Y.-T. Wu, F. Porté-Agel (2015), Modeling turbine wakes and power losses within a wind farm using LES: An application to the Horns Rev offshore wind farm, *Renew Energy*, 75, pp. 945-955
[12] Z. Ti, M. Zhang, Y. Li, K. Wei (2019), Numerical study on the stochastic response of a long-span sea-crossing bridge subjected to extreme nonlinear wave loads, *Eng Struct*, 196, p. 109287
[13] M.P. van der Laan, N.N. Sørensen, P.E. Réthoré, J. Mann, M.C. Kelly, N. Troldborg, *et al.* (2015), An improved k-ε model applied to a wind turbine wake in atmospheric turbulence, Wind Energy, 18, pp. 889-907
[14] A. El Kasmi, C. Masson (2008), An extended k–ε model for turbulent flow through horizontal-axis wind turbines, *J Wind Eng Ind Aerodyn*, 96, pp. 103-122
[15] Maziar Raissi, Alireza Yazdani, and George Em Karniadakis (2018), Hidden Fluid Mechanics: A Navier-Stokes Informed Deep Learning Framework for Assimilating Flow Visualization Data, Division of Applied Mathematics, Brown University
[16] M.Raissi, P.Perdikaris, G.E.Karniadakis (2019), Physics-informed neural networks: A deep learning framework for solving forward and inverse problems involving nonlinear partial differential equations, *Journal of Computational Physics*, 378, pp. 686-707
[17] L. Lu, X. Meng, Zhiping Mao, G. E. Karniadakis (2020), Deepxde: A deep learning library for solving differential equations, Division of Applied Mathematics, Brown University
[18] B.D. Tracey, K. Duraisamy, J.J. Alonso (2015), A machine learning strategy to assist turbulence model development, *53rd AIAA aerospace sciences meeting*, p. 1287
[19] J. Ling, A. Kurzawski, J. Templeton (2016), Reynolds averaged turbulence modelling using deep neural networks with embedded invariance, *J Fluid Mech*, 807, pp. 155-166
[20] A. Biswas, R. Gupta (2009), An artificial neural network based methodology for the prediction of power & torque coefficients of a two bladed airfoil shaped H-rotor, *Open Renew Energy J*, 2, pp. 43-51
[21] A. Biswas, S. Sarkar, R. Gupta (2016), Application of artificial neural network for performance evaluation of vertical axis wind turbine rotor, *Int J Ambient Energy*, 37, pp. 209-218
[22] Z. Ti, X. W. Deng, H. Yang (2020), Wake modelling of wind turbines using machine learning, *Applied Energy*, 257, 114025
[23] N. Dimitrov (2018), Surrogate models for parameterized representation of wake-induced loads in wind farms, *Wind Energy*, 22 (43)

[24] R. Riva, J. Liew, M Friis-Møller, N. Dimitrov, E. Barlas, P.E. Réthoré and A. Beržonskis (2020), Wind farm layout optimization with load constraints using surrogate modelling, *Journal of Physics Conference Series*, 1618, 042035

[25] N. Dimitrov and A. Natarajan (2021), Wind farm set point optimization with surrogate models for load and power output targets, 2018 (1), 012013

[26] Y. Bengio (2012), Practical recommendations for gradient-based training of deep architectures, Cornell University, arXiv:1206.5533

[27] K.A. Kragh, M.H. Hansen (2015), Potential of power gain with improved yaw alignment, *Wind Energy*, 18, pp. 979-989

[28] T.J. Chung (2002), *Computational Fluid Dynamics*, Cambridge Univ. Press.

[29] S. Ioffe, Ch. Szegedy, (2015), Batch normalization: Accelerating Deep Network training by reducing internal covariate shift, arXiv:1502.03167v3

[30] I. Guyon (1977), A scaling law for the validation-set training-set size ratio, *AT&T Bell Laboratories*, Berkeley, California

[31] M. Riedmiller, H. Braun (1993), A Direct Adaptive Method for Faster Backpropagation Learning: The RPROP Algorithm, *International Conference on Neural Networks*

[32] J. Kuo, D. Romero, C.H. Amon (2014), A novel wake interaction model for wind farm layout optimization, *ASME 2014 Intl. Mech. Engin. Congress,* Nov. 14-20, Monteral, Canada

[33] T. Gocmen, P. Laan, P.E. Rethore, A.P. Diaz, G.Ch. Larsen, S. Ott (2016), Wind turbine wake models developed at the technical university of Denmark: A review, *Renewable & Sustainable Energy Reviews*, 60, pp. 752-769

[34] P.M. Gebraad, F.W. Teeuwisse, J.W. van Wingerden, P.A. Fleming, S.D. Ruden, J.R. Marden, L.Y. Pao (2016), Wind plant power optimization through yaw control using a parametric model for wake effects – A CFD simulation study, *Wind Energy,* 19, pp. 95-114

[35] M. Samorani (2010), The wind farm layout optimization problem, DOI: 10.1007 / 978-3-642-41080-2_2

[36] D.E. Goldberg DE (1989), *Genetic Algorithms in Search, Optimization and Machine Learning*, Addison-Wesley Longman Publishing Co., Inc., Boston

[37] E.G.A. Antonini, D.A. Romero, C.H. Amon (2020), Optimal design of wind farms in complex terrains using computational fluid dynamics and adjoint methods, *Applied Energy*, 261, 114426.

[38] R.N. King, K. Dykes, P. Graf, P.E. Hamlingon (2016), Adjoint optimization of wind plant layouts, *Wind Energy Science Discussions,* doi:10.5194/wes-2016-25, 2016

[39] S.W. Funke, R.E. Farrell, MD. Piggott (2013), Tidal turbine array optimization using the adjoint approach, *Renewable Energy*, 63, pp. 658-673

[40] S. Dongran, J. Yang, Y. Liu, M. Su, A. Liu, Y.H., Joo (2017), Wind direction prediction for yaw control of wind turbines, Intl., J. of Control Automation and Systems, 15, pp. 1-9

[41] A.S. Dar, L. von Bremen (2019), Short-Term Forecasting of Wake-Induced Fluctuations in Offshore Wind Farms, *Energies*, 12, 2833; doi:10.3390/en12142833

Journal of Physics: Conference Series **2151** (2022) 012012 doi:10.1088/1742-6596/2151/1/012012

Deriving an experimental and analytical relation between the core and fiber temperatures of a 3P XLPE cable

Manly Callewaert / Roel Vanthillo / Christian Fojtek (Marlinks)
Bart Mampaey (Elia)

manly.callewaert@marlinks.com

Every submarine power cable contains a fiber-optic assembly. Most windfarms have a distributed temperature sensing (DTS) device for continuous temperature monitoring. Using real-time thermal rating (RTTR) the performance of subsea cables can be increased or a less expensive cable can be installed.

The experiments aimed to establish the relation between the fiber and core (or conductor) temperatures in the cable. Fiber temperature was obtained using a DTS interrogator; thermocouples installed in the conductors yielded the core temperature data. The cable used in this test was a 20 meter-long 1800 mm^2, 1000 A export cable.

These experiments validated a new RTTR model proposed by Marlinks: a thermal RC-ladder model based on finite-element (FEM) calculations. Better suited to the complexity of the spacer geometry and existence of the fiber node, the Marlinks-developed model is considered to be more accurate than the industry standard.

1. Objectives

Reaching conductor temperatures of 90°C or higher will damage the XLPE layer which can render the whole cable useless. This is why some wind farms have RTTR modules to protect their assets. Most of these modules use the IEC industry standard to calculate the core temperature, however we find this method lacking as it does not take into account the increased geometrical complexity of modern cables due to the presence of spacers and a fiber node inside the cable (Figure 1).

This is why, in this paper, we propose and experimentally validate a new thermal RC-ladder RTTR model based on FEM calculations. The FEM calculations themselves have high predictive power, however, they are also too computationally demanding for real-time applications. The proposed new model tackles this issue. We will also argue why the IEC industry standard [1] is found imperfect for the purpose of RTTR.

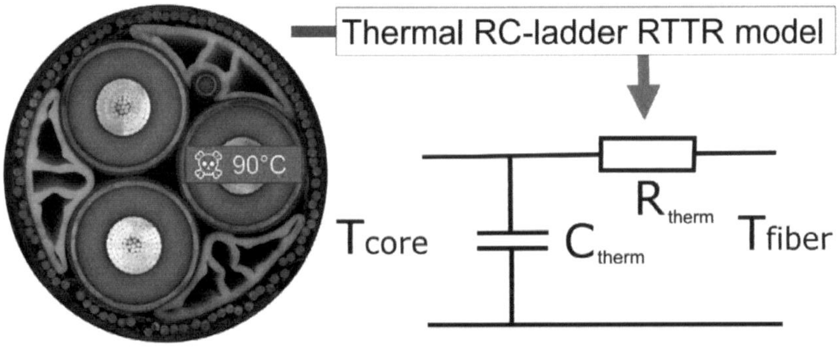

Figure 1 A method is proposed to translate a complex cable geometry into a simple model. Left: picture of a modern offshore high voltage export cable cross-section. Right: thermal RC-ladder model for RTTR modeling.

2. Methods

Prior to setting up the real-life experiment, the heat flows inside the cable and in the sand bed were described in the highest possible detail using an FEM model calculation. Next, an alternative and computationally less intensive method was proposed [2].

The new method uses a numerical iterative RC-ladder model (Figure 1) to calculate the core temperature from the measured fiber temperature obtained by a DTS interrogator. The thermal resistance *R* and capacitance *C* were established by fitting a dynamic FEM calculation [3-5]. To validate this approach, an experiment is set up with thermocouples measuring the actual conductor temperature.

2.1. FEM modeling

The cable as described in the datasheet [6] was FEM-modeled. A detailed CAD drawing of the cable was made. The cable datasheet also provided the materials used in the cable's construction. The thermal properties of the individual cable parts where found in a database [7]. The properties of the water-saturated sand were measured in the University of Ghent's geotechnics laboratory. The sand bed is simplified as an 80 cm-wide square. Figure 2 shows the effect of 1500 A applied for 24 hours. This explorative calculation was done prior to constructing the box. This calculation clearly shows that the dimensions of the box are chosen so that the heat does not reach the wall, and therefore that, for the purpose of the experiment, the cable can be considered buried in a semi-infinite body.

The IEC standard, which was developed before the expansion of offshore wind and before the widespread adoption of fiber-optic cables and DTS equipment, does not allow calculating thermal R or C to a fiber node such as the one present in all modern submarine cables; it assumes the temperatures to be measured at the cable wall. To use the standard with DTS-based temperatures from the fiber node instead, the standard should be used to calculate cylindrical shell formulas up to the cable sheath. While the fiber is located close to the sheath in most cables, this method lacks the accuracy of the FEM calculation which is better able to deal with the highly complex spacer geometry inside the cable.

Based on a time-dependent simulation (Figure 3), these thermal RC-values can be fitted. The load profile used in this numerical calculation was equal to the ones used in the experiments (Figure 7, left panel). The resulting values are shown in Table 1.

R [K/W]	0.0817
C [J/K]	27373

Table 1 RC-values found by fitting a dynamic FEM calculation

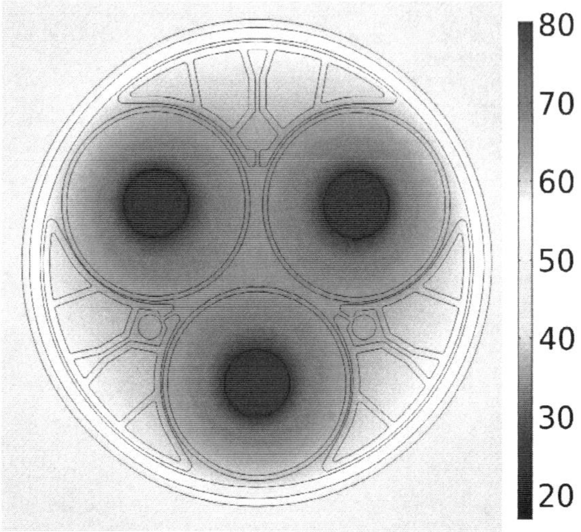

Figure 2 An FEM model is built using the material properties and cable geometry obtained from the cable datasheet. The figure shows the results of 24h constant load (1500 A) calculation, with core temperatures reaching 80°C (zoomed in).

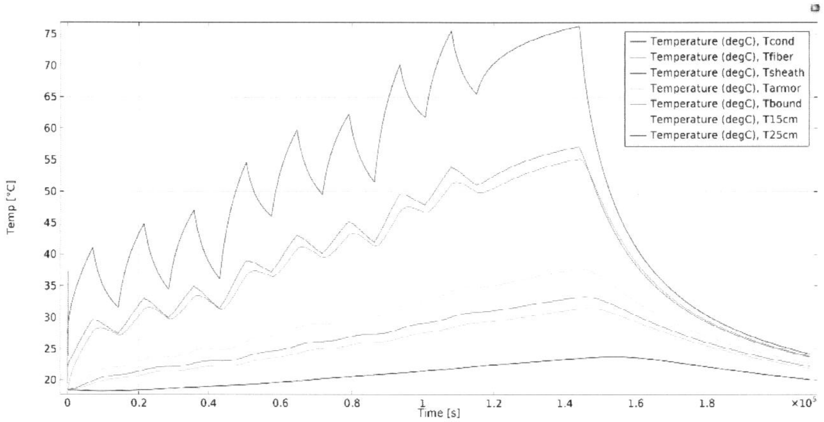

Figure 3 Result of a dynamic load profile FEM simulation prior to the experiment. Temperature versus time plots from important nodes inside the cable.

2.2. Experimental set-up

The cable used for this test is a 20 m-long, 1800 mm², 1000 A export cable. We buried it in the center of a 80 cm-wide square duct (Figure 4). The experiment (Figure 5) consists of 3 current transformers and a star point. There were 9 thermocouples installed inside the cores, i.e. 3 in each phase. Thermocouples were installed at various distances on 3 wooden frames to monitor the heat propagating inside the box.

The cable's nominal load was 1000 A, however in order to reach sufficiently high temperatures in the experimental window of a few days, much higher currents of 1500-1800 A were induced into each of the three conductors with current injection transformers (Figure 7). The three phases were connected in star with aluminum cable terminations (Figure 6). The soil and armor were grounded together. Power capacitators were used to increase the cos phi and to avoid power supply failures.

The dimensions of the duct were chosen so that the heat would not reach the boundaries within a 24-hour period of constant heating. To remove the heat and prepare the setup for another experiment, a drainage system was installed underneath the sand bed consisting of drain pipes and lava stone.

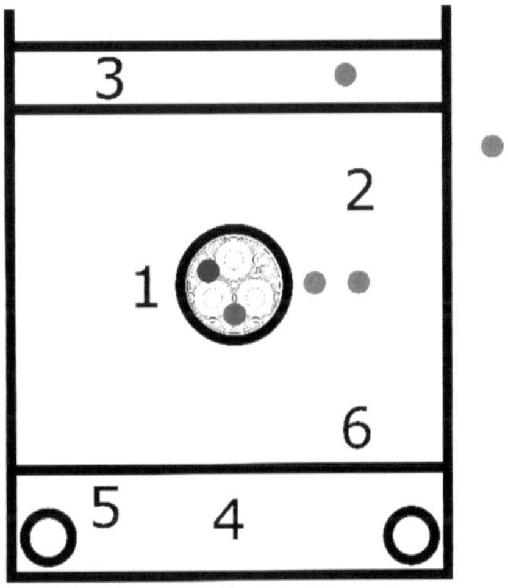

Figure 4 Cross-sectional drawing of the box. Red dots: thermocouples outside the cable

1) cable 2) sand 3) water level 4) lava stone layer 5) drainpipe 6) canvas

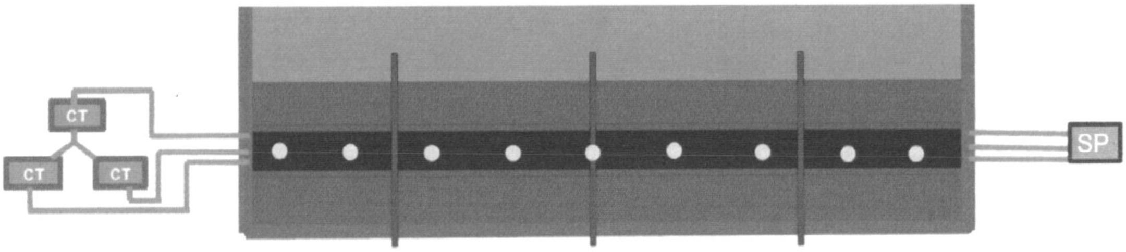

Figure 5 Schematic diagram of the experiment. Yellow dots: thermocouple inside the conductor, Red lines: positions of wooden frames with thermocouples, CT: current transformer, SP: star point.

3. Results

3.1. Construction

The offshore high-voltage cable was installed when the box was constructed and half filled with sand (Figure 6, top left). Three wooden frames (Figure 4) with thermocouples were installed to monitor the heat propagation inside the box. Based on the lay length of the cable, nine equidistant spaced holes were drilled in the cores to install thermocouples (Figure 6, bottom left and Figure 5, yellow dots). The star point was made by short circuiting the three phases with thick aluminum plates.

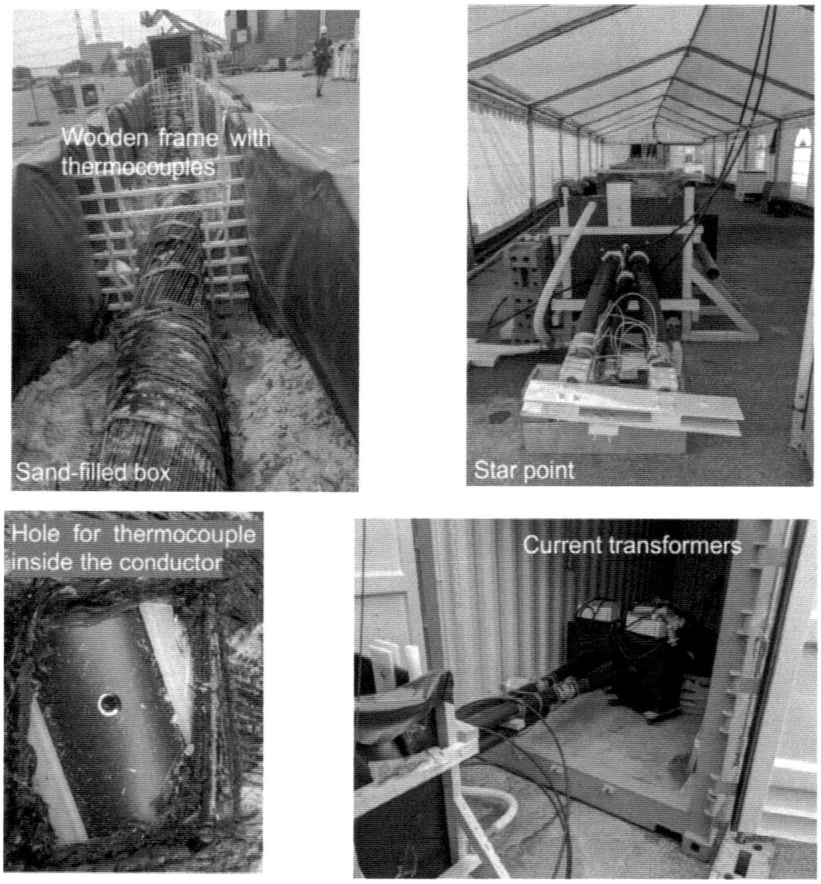

Figure 6 Overview pictures of the set-up

3.2. Experiments

We performed a first experiment with 24 hours of constant load, and a second with a dynamic load profile. It is important to note that much more current was applied than the nominal 1000 A. In the first experiment, the load was constantly 1500 A; in the dynamic-load case the maximum load briefly climbed to 1800 A.

In the first phase of the experiment, a constant current of 1500 A was applied for 24 hours. Thereafter, the current was switched off for 18 hours. In the second phase, a dynamic-load profile was applied to the cable for 36 hours after which the cable was allowed to cool for 18 hours (Figure 7). Ultimately, we removed the heat in preparation for a new experiment by opening the drain taps and flushing with fresh water.

Figure 8 shows graphs of the experiments. The red line indicates the measured DTS temperature, the black line shows the measured core temperature, and the dashed line is the prediction made using our RC-ladder RTTR model based on the FEM calculation. It should be abundantly clear that this is a very good fit.

A nominal load of 1000 A proved to be very conservative as the cable sustained a continuous load of 1500 A without reaching the critical temperature of 90°C. The underpinning IEC norm is considered to be equally conservative and inadequate. The FEM-based method is demonstrably the better approach in terms of RMS prediction errors. This also means that the end-user needs not build in any additional safety factors when evaluating core temperatures as these are already incorporated in the IEC norm.

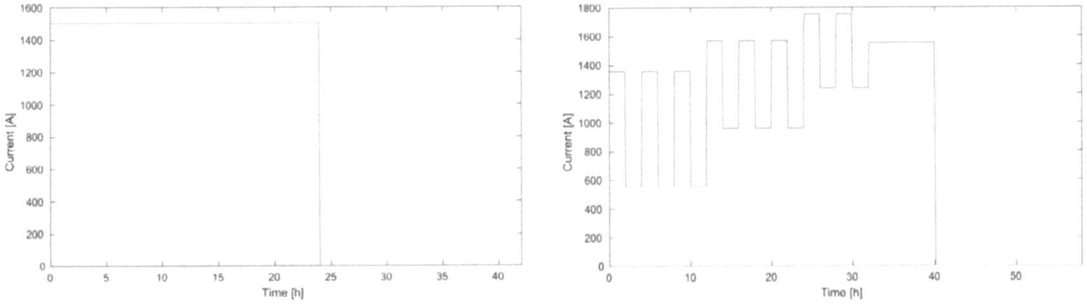

Figure 7 Current profiles applied during the experiment: Left: 24h constant current, Right: synthetic dynamic load.

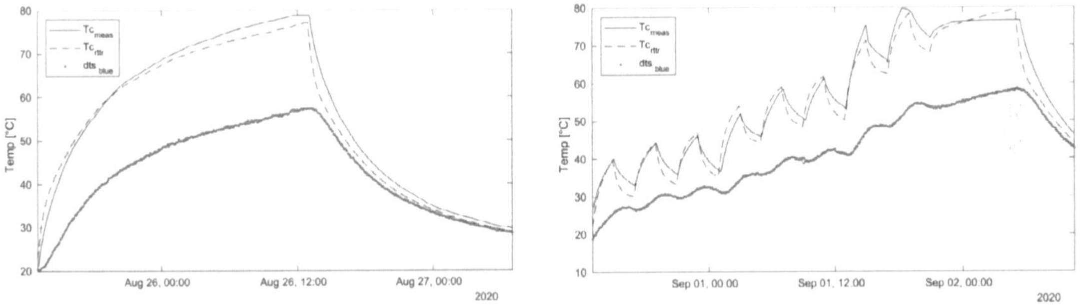

Figure 8 Comparison of the measured core temperatures and those predicted by our RTTR model based on the measured DTS temperature.

4. Conclusion

Having accurate and real-time insights in core temperature allows end users to maximize power output and rapidly react to rising core temperatures without risking cable failure.

The experiments laid out in this paper validate the proposed new approach's predictive power for providing such accurate and real-time insights. The RC-ladder values derived from the FEM calculation's results are found to be very accurate, and they far less computationally intensive than the FEM calculation itself.

It was also shown that the proposed RTTR model enables cables to transport much higher currents than the maximums dictated by the in our view outdated industry standard.

BIBLIOGRAPHY

[1] IEC 60853-1, "Calculation of the cyclic and emergency current rating of cables"
[2] M. Diaz-Aguiló, F. De León, S. Jazebi, M. Terracciano, "Ladder-Type Soil Model for Dynamic Thermal Rating of Underground Power Cables" (IEEE Power and Energy Technology Systems Journal, 10.1109/JPETS.2014.2365017)
[3] S. Strum, A. Küchler, J. Paulus, "3D-FEM modelling of losses in armoured submarine power cables and comparison with measurements" (Cigré session 48, B1-215, Paris 2020)
[4] S. Sturm, J. Paulus, K.-L. Abken, F. Berger, "Estimating the losses in three-core submarine power cables using 2D and 3D FEA simulations" (Jicable, F2.23, Paris 2015)
[5] 2021-10-13 Meeting notes - BE B1 Insulated Cables - CIGRE Groups
[6] Hellenic Cables S.A. (Datasheet: MOG-HEC-EG-RPT-00001)
[7] http://www.matweb.com

 2151 (2022) 012013 doi:10.1088/1742-6596/2151/1/012013

Offshore Wind-to-Hydrogen Production Plant Integrated with an Innovative Hydro-Pneumatic Energy Storage Device

J Settino[1], R N Farrugia[1], D Buhagiar[2] and T Sant[3]

[1] Institute for Sustainable Energy, University of Malta, Marsaxlokk MXK1531, Malta.
[2] FLASC B.V., Delft, The Netherlands.
[3] Department of Mechanical Engineering, University of Malta, Msida MSD2080, Malta.

robert.n.farrugia@um.edu.mt

Abstract. One of the main technical challenges of wind-to-hydrogen production plants is to couple intermittent and variable renewable power sources, such as wind turbines, with an electrolyser. Most hydrogen production concepts rely on electrical storage to smoothen the power input to the electrolyser. In this study, the use of a hydro-pneumatic energy storage system is proposed as an interface between the green, fluctuating electricity supply and the electrolyser. The performance of the proposed solution is analysed and compared to that of a conventional offshore wind-to-hydrogen production plant in order to identify potential advantages and drawbacks. The results of the simulations show that the FLASC energy storage system provides a stabilized, yet green power supply. Short-term fluctuations of the power input to the electrolyser can be avoided. A 70% reduction of the On/Off cycles can be obtained. Moreover, if compared to an offshore wind-to-hydrogen production plant with no storage there is no substantial difference in terms of hydrogen production observed over the analysed period of one year in spite of a 70% round-trip efficiency of the energy storage device.

1. Introduction

Hydrogen is expected to become a key energy carrier [1]. According to IRENA [2], the hydrogen demand will be of over 160 million metric tons by 2050. Currently, up to 96% of the overall hydrogen is produced using fossil fuels (coal, oil and natural gas), and only 4% is produced by electrolysis, as underlined by Kothari et al. [3]. Electrolysis is a process that allows for splitting of water molecules into hydrogen and oxygen using electricity. In contrast to steam reforming, gasification and partial oxidation of fossil fuels, electrolysis can be considered a green technology if renewable energy is used instead to power the electrolyser [4].

Coupling intermittent power sources, such as wind turbines, with an electrolyser represents one of the main technical challenges of green hydrogen production plants [5, 6]. Most hydrogen production concepts rely on electrical storage to smoothen the power input to the electrolyser [7]. The study performed by Gahleitner [8] shows that Lead-acid batteries were used in 53% of the analysed power-to-gas plants when the renewable energy generator was directly connected to the electrolyser. These smoothen the power output of the generator, thus minimising the On/Off cycling of the electrolyser. Even though storage devices mitigate the problems associated with the stochastic nature of the wind resource, conventional onshore energy storage technologies are usually not ideal for large-scale offshore applications. Battery storage systems introduce significant safety hazards in a highly-volatile hydrogen production environment and take up precious floating platform deck space if co-located near offshore RE systems. Moreover, batteries have to be replaced every 5 to 7 years.

In this study, an innovative approach has been proposed: using a Floating Liquid-piston Accumulator using Seawater under Compression (FLASC) energy storage system (ESS) as an interface between the green, fluctuating electricity supply and the electrolyser.

The FLASC system is a type of Hydro-Pneumatic Energy Storage (HPES) technology [9], that consists of two chambers: the upper floating air chamber (A) and the liquid piston chamber (B) that is anchored at the seabed [10]. The two chambers are inter-connected via a pneumatic umbilical. A schematic representation of the FLASC system is shown in Figure 1. A small-scale prototype of the FLASC technology was deployed in Malta in 2017 (Figure 2). The testing campaign served to demonstrate the working principle, measure the thermodynamic response and validate the numerical models required when designing larger embodiments [11, 12]. The technology is presently being up-scaled and industrialised through a number of ongoing initiatives.

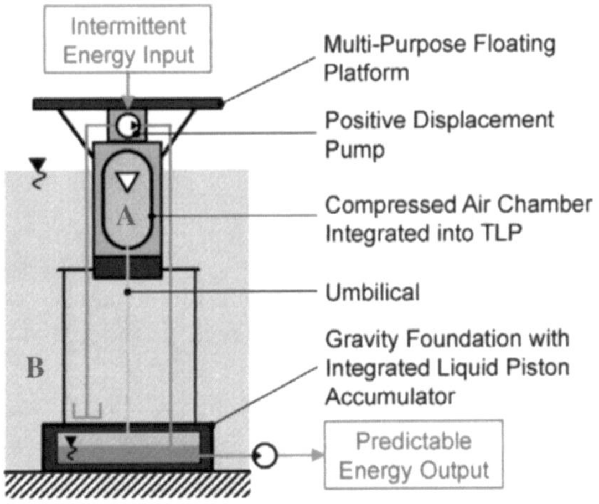

Figure 1. Schematic representation of the FLASC energy storage system. Source: [6]

Figure 2. Small-scale prototype of the FLASC system deployed in Malta. Source: [11]

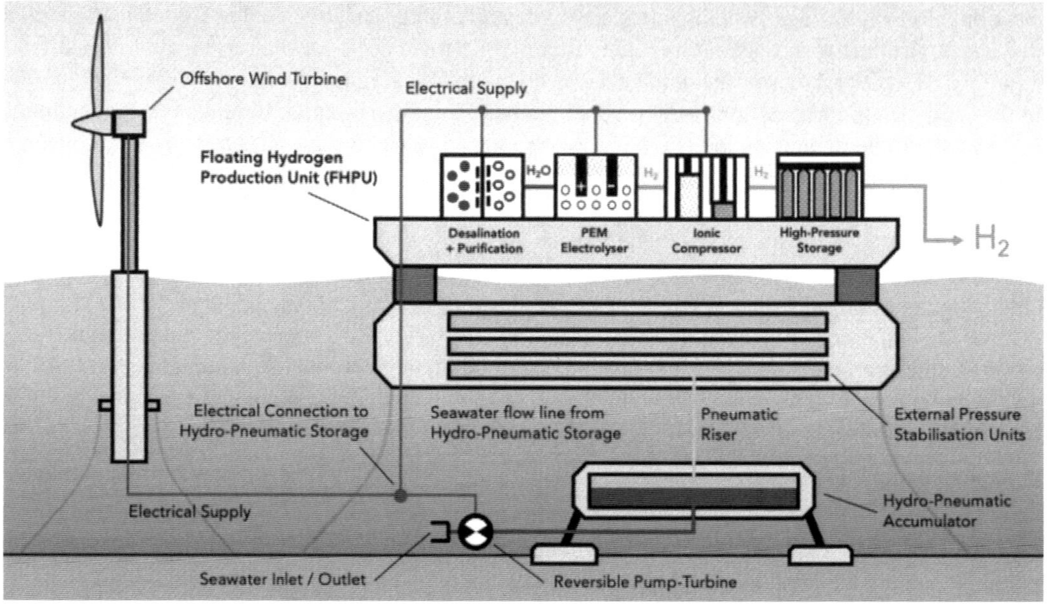

Figure 3. Innovative wind-to hydrogen plant interfaced with a FLASC-integrated system

In the initial discharged state, both chambers are filled with air at a pre-set pressure. During charging, the electricity supplied by the renewable source is used to pump seawater into the seabed component by means of a reversible pump / turbine; at this stage operating in pump mode. Water pumped into the lower chamber decreases the internal air volume across both chambers as the air also flows through the umbilical and up into the upper floating chamber, A. The system is fully-charged when the prescribed limit to the water capacity of chamber B is reached. When no wind power is available, the compressed air inside the hydro-pneumatic storage vessels is allowed to expand, expelling seawater at high pressure. The pressurised seawater drives the reversible pump-turbine, now operating in turbine mode. If coupled to an intermittent renewable source such as the wind turbine, the FLASC system can supply a stable supply of green power. A schematic representation of the system concept is provided in Figure 3.

The electricity generated by the wind turbine coupled with a FLASC system is used to power the electrolyser, the ionic compressor and all the auxiliaries. The performance of the proposed solution is analysed and compared to that of a conventional wind-to-hydrogen production plant without storage.

2. Methodology

In order to compare the performance of the conventional wind-to-hydrogen production plant against that of the proposed innovative concept, simulations of the two processes were carried out. For the sake of comparison, the same process layout has been considered in both scenarios being analysed. Seawater is fed to the desalination unit, consisting of a filtration and reverse osmosis (RO) system. Afterwards, the permeate water is passed on to the proton exchange membrane (PEM) electrolyser to be converted to hydrogen, which is then pressurized through the ionic compressor.

In both scenarios, a typical 6 MW wind turbine has been considered. The characteristic power curve of the wind turbine is provided in Figure 4. For the analysis, one year of data (from July 2005 to June 2006) from the Egmond aan Zee offshore wind farm site [13], with an averaging period of 10 minutes, was considered. Mathematical models of the different system components were developed and validated based on scientific data available in literature or provided by manufacturers. Detailed models of the desalination unit, PEM electrolyser and ionic compressor were implemented in Matlab 2020a [14] and are described in the next paragraphs. Additional components were also modelled, such as pressure exchangers, pumps and converters. All the subsystems were integrated to simulate the overall wind-to-hydrogen production plant.

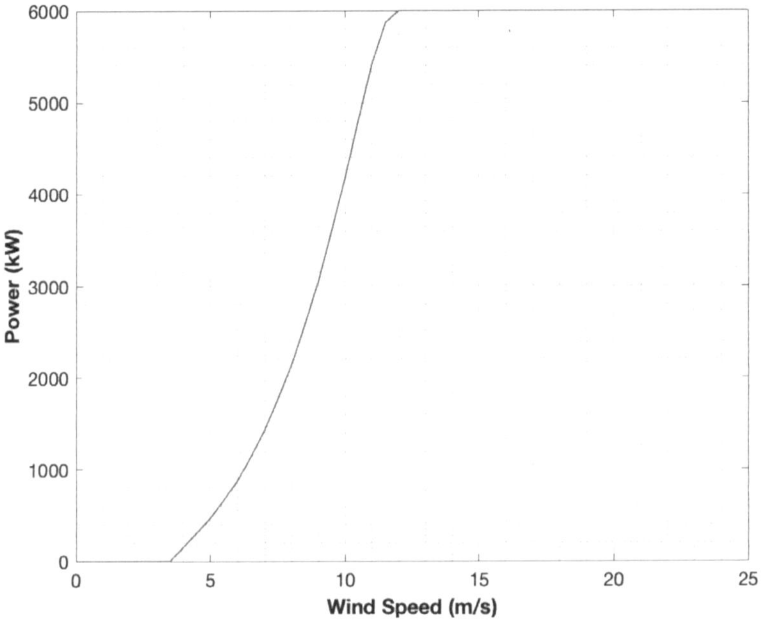

Figure 4. Power Curve of a generic 6 MW wind turbine

2.1. Reverse Osmosis (RO)

In the reverse osmosis (RO) process, a semi-permeable membrane is used to separate seawater flow, Q_f, into two streams: a concentrate solution (brine) flow, Q_b, and a purified water (permeate) flow, Q_p. Several models have been developed and proposed by a number of authors to determine the performance of a reverse osmosis system. They can be classified in two main categories: solution-diffusion models and pore-flow models [15, 16]. In the solution diffusion models, it is assumed that the transport through the membrane occurs only by diffusion and that the permeants dissolve in the membrane and then diffuse. On the other hand, the pore-flow models consider the presence of tiny pores on the membrane and assume that the permeants are transported by pressure-driven convective flow. A mathematical model for the Reverse Osmosis process based on the solution-diffusion approach, has been developed. The water flux (J_W) and the salt flux (J_S) through the membrane can be determined based on the solution-diffusion mass transfer model such that:

$$J_W = A[(p_b - p_P) - (\pi_W - \pi_P)] \tag{1}$$

$$J_s = B(C_w - C_P) \tag{2}$$

where A and B are the pure water permeability and the solute permeability constants respectively. The variable p represents the pressure; with the subscript p referring to the permeate and the subscript b indicating the brine. The concentration of solute is C while the symbol π represents the osmotic pressure that is a function of the concentration.

The main input variables of the model are: the characteristics of the membrane provided by the manufacturer (variables A and B), the seawater solute concentration, pressure and temperature. The model provides the following outputs: the permeate flow rate (Q_p), the brine flow rate (Q_b) and their concentrations.

The model was tested by varying the input parameters to cover a wide range of operation: from low seawater salinity (15,000 ppm) to a high solute concentration (45,000 ppm) and was validated based on data available in literature [17, 18].

2.2. The PEM Electrolyser

The models proposed in literature to investigate the performance of a PEM Electrolyser can be classified depending on: the physical domain involved (electrochemical, thermal and mass transfer models), the modelling approach (analytical or empirical) and on the scale (single cell, stack or overall system) of the unit [19]. All models aim at describing electrolyser behaviour. The performance of an electrolyser is summarized by its polarisation curve, which shows the relationship between the voltage and the current density.

In this study, the polarization curve was determined based on physical laws and equations. The cell voltage was obtained considering the reversible potential (E_{rev}), the activation overvoltage ($E_{act,an}, E_{act,cat}$), the ohmic overvoltage (E_{ohm}) and the diffusion overvoltage ($E_{diff,an}, E_{diff,cat}$), as described in detail by Abdin et al. [20] as follows:

$$E_{cell} = E_{rev} + E_{act,an} + E_{act,ca} + E_{ohm} + E_{diff,an} + E_{diff,cat} \tag{3}$$

The hydrogen production rate was estimated based on Faraday's Law [21].

$$\dot{n}_{H2} = \eta_F \frac{n_c I}{zF} \tag{4}$$

where $\dot{n}_{H2}$ is molar flow of hydrogen in mol/s, I is the current in A, F is the Faraday Constant (96485 C/mol), z is the number of electrons transferred for each mol of hydrogen produced (in this case, it is 2), n_c is number of electrolyser cells and η_F is the Faraday efficiency (usually considered equal to 1). The results of the PEM Electrolyser model were validated and are in agreement with published material [22, 23].

2.3. The Ionic Compressor (IC)

The ionic compressor represents an interesting alternative to conventional multi-stage compressors. It was patented by the Linde Group in 2006 [24]. Since then, it has been commercialised and is widely used for hydrogen compression. According to the information provided by the manufacturer, a nearly isothermal compression process occurs and an efficiency of almost 100% is reached. In this study, a model of the ionic compressor was developed considering the real gas behaviour of hydrogen at high pressures. The ideal gas law is suitable to represent the behaviour of hydrogen up to 10 MPa, but this leads to an error of 44% when used for pressures of 70 MPa [25]. At high pressures, hydrogen shows a non-linear correlation between density and gas pressure. In order to consider this effect, the Redlich–Kwong equation was implemented [26]. Hence, a more accurate prediction of the work required to compress the hydrogen can be obtained, so that:

$$W = R\,T \ln\left(\frac{v_2 - b}{v_1 - b}\right) + \frac{a}{b\sqrt{T}} \ln\left\{\frac{(v_2 + b)\,v_1}{(v_1 + b)\,v_2}\right\} \tag{5}$$

where R is the gas constant, v the molar volume ($m^3\ mol^{-1}$), T is the temperature (K) and the subscripts 1 and 2 refer to the initial and final states of the gas respectively.

3. Results

Using the software tool developed in Matlab 2020a, simulations were carried out to analyse the performance of both a conventional wind-to-hydrogen production plant and the proposed concept with integrated storage. The main results are summarised in Table 1.

Table 1. Main simulation results.

Wind-to-Hydrogen Plant without storage	Wind-to-Hydrogen Plant with storage
Generated Electricity: $2.65 \cdot 10^7$ kWh	Generated Electricity: $2.65 \cdot 10^7$ kWh
H_2 Production: $4.17 \cdot 10^6$ Nm^3	H_2 Production: $4.17 \cdot 10^6$ Nm^3
Energy Losses: $2.828 \cdot 10^5$ kWh (1.1 %)	Energy Saved: $1.98 \cdot 10^5$ kWh (0.75 %)
Minimum Power Required: 512.181 kW	Power at 1 A/cm^2: 2631.5 kW
-	Additional H_2 Production: $3.4 \cdot 10^4$ Nm^3 (0.82 %) (455.256 Nm^3/h * 75.24 h)
Number of ON/OFF cycles: 543	Number of ON/OFF cycles: 162 (70% reduction)

The electricity generated by the wind turbine is the same in both scenarios, but the analysis shows that about 1.1% of the generated electricity is lost in conventional wind-to-hydrogen plants. A minimum power of 512.181 kW is required to operate the electrolyser and all the auxiliaries. Hence, all the power below this threshold value is lost if no storage device is used.

In the analysis of the wind-to-hydrogen plant with integrated storage, it was assumed that the energy lost can be exploited. The energy saved is determined considering the energy lost in the conventional system with no storage and considering a round-trip efficiency (η_{rt}) of 0.7 for the FLASC-integrated system, such that:

$$E_{saved} = E_{lost} \cdot \eta_{rt} \tag{6}$$

The additional hydrogen production was estimated considering that the energy recovered as a result of the storage device was used to power the overall hydrogen production plant. Since a power of 2631.5 kW is required to operate the electrolyser at a current density of 1 A/cm^2 with a corresponding hydrogen production rate of 455.256 Nm^3/h, the number of additional hours during which the system is in operation has been defined as the ratio between the saved energy and the power required. The additional hydrogen produced can be estimated by multiplying the hydrogen production rate and the number of hours.

It is worth noting that the additional hydrogen produced represents a marginal increase if compared to the overall hydrogen production (at below 1%). This is important to underline since, despite the inclusion of a storage device with a round-trip efficiency of 70%, no reduction in H_2 production was observed. The energy lost due to the non-ideal efficiency of the ESS is compensated for by the energy savings. The latter includes:

- the energy below the minimum threshold value required to operate the whole plant;
- the energy recovered thanks to the reduced number of ON/OFF cycles. A period of 20 minutes is necessary to start a PEM electrolyser after a shut down.
- The number of ON/OFF cycles was determined for a storage system sized to maintain a fixed power output for 2 hours. Compared to the conventional 'no storage' option, a 70% reduction in cycles could be achieved. This is due to the FLASC-integrated system that enables the

averaging of the overall power output over a determined period of time. Hence, as shown in Figure 5, short-term fluctuations of the power input to the electrolyser can be avoided. In Figure 5, the electricity generated by a typical 6 MW wind turbine is represented in blue while the power output provided by the wind turbine coupled with the FLASC-integrated system is shown in red.

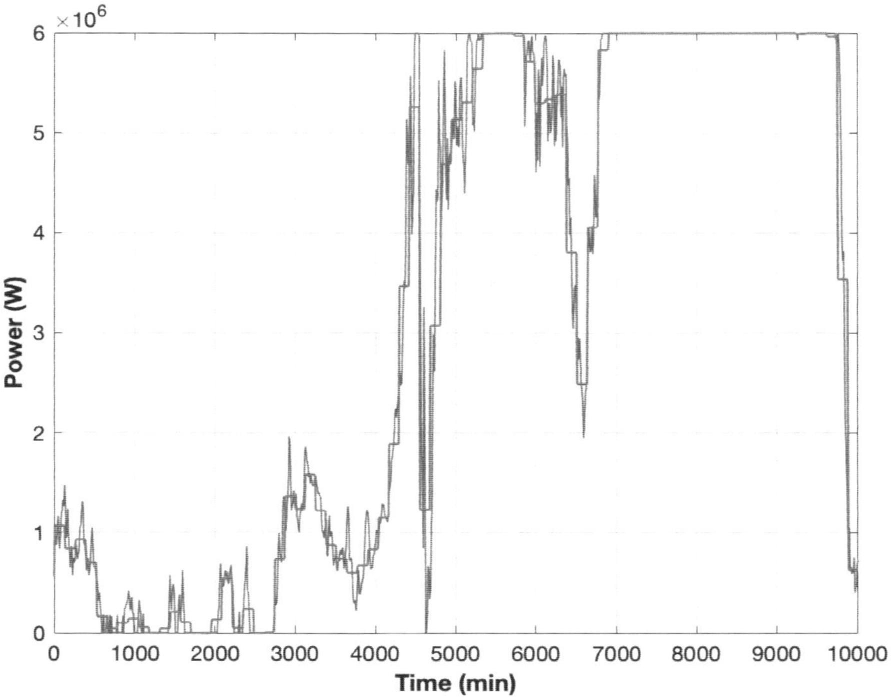

Figure 5. Comparison between the electricity generated by a 6 MW wind turbine and the output of a wind turbine coupled with the FLASC-integrated system having a storage period of 2 hours

4. Conclusions

The analysis performed shows that the proposed concept allows for reducing short-term power fluctuations and the number of electrolyser ON/OFF cycles by some 70%. This is expected to increase the electrolyser lifetime substantially. Moreover, despite the FLASC system's 70% round-trip efficiency, no reduction in H_2 production was observed. The benefit of including a storage system is evident. Nevertheless, the proposed solution provides additional benefits also compared to a wind-to-hydrogen production system using electrical storage (batteries) in that no additional space is required on an offshore platform and that there is no fire hazard given that none of the materials used in the FLASC ESS is flammable. Moreover, compared to batteries that have to be replaced, the proposed solution is designed for a longer lifetime of more than 30 years of continuous operation, i.e. similar to the wind turbine and electrolyser expected lifetime. Additionally, the FLASC system utilises components whose performance characteristics are already proven and well-understood.

Acknowledgments & Collaborations

Project supported through the Maritime Seed Award (MarSA) 2019 funded by Transport Malta (formerly Malta Marittima). Project carried out in collaboration with FLASC B.V., Delft, The Netherlands.

References

[1] Rosen M A and Koohi-Fayegh S 2016 The prospects for hydrogen as an energy carrier: an overview of hydrogen energy and hydrogen energy systems *Energy Ecology and Environment* **1**, pp 10-29.
[2] International Renewable Energy Agency (IRENA*), Green Hydrogen Cost Reduction Scaling Up Electrolysers to Meet the 1.5°C Climate Goal*, 2020.
[3] Kothari R, Buddhi D, and Sawhney R L 2008 Comparison of environmental and economic aspects of various hydrogen production methods *Renew. Sustain. Energy Rev* **12** pp 553–563.
[4] Kakoulaki G, Kougias I, Taylor N, Dolci F, Moya J and Jager-Waldau A 2021 Green hydrogen in Europe - A regional assessment: Substituting existing production with electrolysis powered by renewables *Energy Convers. Manag.* **228** p 113649
[5] Egeland-Eriksen T, Hajizadeh A, and Sartori S 2021, "Hydrogen-based systems for integration of renewable energy in power systems: Achievements and perspectives," *Int. J. Hydrogen Energy*, vol. 46, no. 63, pp. 31963–31983, doi: 10.1016/J.IJHYDENE.2021.06.218
[6] Weiß A, Siebel A, Bernt M, Shen T H, Tileli V, and Gasteiger H A 2019, "Impact of Intermittent Operation on Lifetime and Performance of a PEM Water Electrolyzer," *J. Electrochem. Soc.*, vol. 166, no. 8, pp. F487–F497, doi: 10.1149/2.0421908JES/XML
[7] Sood S *et al.* 2020, "Generic dynamical model of PEM electrolyser under intermittent sources," *Energies*, vol. 13, no. 24, pp. 1–34, doi: 10.3390/en13246556
[8] Gahleitner G 2013 Hydrogen from renewable electricity: An international review of power-to-gas pilot plants for stationary applications *Int. J. Hydrogen Energy* **38** pp 2039–2061.
[9] Buhagiar D, Sant T, and Farrugia R N 2021, "Hydro-Pneumatic Energy Storage," *Ref. Modul. Earth Syst. Environ. Sci.*, doi: 10.1016/B978-0-12-819723-3.00077-9.
[10] Buhagiar D and Sant T 2017 Modelling of a novel hydro-pneumatic accumulator for large-scale offshore energy storage applications *J. Energy Storage* **14** pp 283–294.
[11] Buhagiar D, Sant T., Farrugia R N, Aquilina L, Farrugia D, and Strati F M 2019, "Small-scale Experimental Testing of a Novel Marine Floating Platform with Integrated Hydro-pneumatic Energy Storage," *J. Energy Storage*, vol. 24, doi: 10.1016/j.est.2019.100774.
[12] Buhagiar D, Sant T, and Farrugia R N 2019, "Marine testing of a small-scale prototype of the FLASC offshore energy storage system," doi: 10.1109/OSES.2019.8867322.
[13] NoordzeeWind, Wind Data from the offshore wind farm. [Online, Accessed on: 20 October 2020]. Available: https://www.noordzeewind.nl/nl_nl/kennis/meteogegevens.html
[14] Matlab 2020a. https://www.mathworks.com/products/new_products/release2020a.html. [Online, Accessed on: 20 October 2020].
[15] Lu Y, Liao A and Hu Y 2012 The design of reverse osmosis systems with multiple-feed and multiple-product *Desalination* **307** pp 42–50.
[16] Baker W R 2004 *Membrane technology and applications* John Wiley and Sons, Ltd.
[17] Marcovecchio M G, Aguirre P A and Scenna N J 2005 Global optimal design of reverse osmosis networks for seawater desalination: modeling and algorithm *Desalination* **184** pp 259–271.
[18] Sassi K M and Mujtaba I M 2012 Effective design of reverse osmosis based desalination process considering wide range of salinity and seawater temperature *Desalination* **306** pp 8–16.
[19] Olivier P, Bourasseau C, Bouamama Pr B 2017 Low-temperature electrolysis system modelling: A review *Renew. Sustain. Energy Rev.* **78** pp 280–300.
[20] Abdin Z, Webb C J and Gray E MacA 2015 Modelling and simulation of a proton exchange membrane (PEM) electrolyser cell *International journal of hydrogen energy* **40** pp 13243-13257.
[21] Görgün H *2006* Dynamic modelling of a proton exchange membrane (PEM) electrolyzer *International Journal of Hydrogen Energy* **31** pp 29–38.
[22] Marangio F, Santarelli M and Calì M 2009 Theoretical model and experimental analysis of a high pressure PEM water electrolyser for hydrogen production *International journal of hydrogen energy* **34** pp 1143–1158.
[23] Han B, Steen III S M, Mo J and Zhang F 2015 Electrochemical performance modeling of a proton

exchange membrane electrolyzer cell for hydrogen energy *International journal of Hydrogen Energy* **40** pp 7006-7016.
[24] https://www.linde-engineering.com/en/images/DS_Ionic%20Fueler%2090_tcm19-523717.pdf
[25] Stolten D 2010 *Hydrogen and Fuel Cells: Fundamentals, Technologies and Applications* John Wiley & Sons.
[26] Annamalai K, Puri I K, Jog M A 2010 *Advanced Thermodynamics Engineering* CRC Press.

Journal of Physics: Conference Series **2151** (2022) 012014 doi:10.1088/1742-6596/2151/1/012014

How do Humans decide under Wind Power Forecast Uncertainty – an IEA Wind Task 36 *Probabilistic Forecast Games and Experiments* initiative

Corinna Möhrlen[1], **Gregor Giebel**[2], **Ricardo J. Bessa**[3], **Nadine Fleischhut**[4]

[1]WEPROG, Assens, Denmark
[2]Technical University of Denmark (DTU), Department of Wind Energy, Roskilde, Denmark
[3]INESC Technology and Science (INESC TEC), Porto, Portugal
[4]Max Planck Institute for Human Development Berlin, Germany

E-mail: [1]`com@weprog.com (corresponding author)`

Abstract. The need to take into account and explicitly model forecast uncertainty is today at the heart of many scientific and applied enterprises. For instance, the ever-increasing accuracy of weather forecasts has been driven by the development of ensemble forecasts, where a large number of forecasts are generated either by generating forecasts from different models or by repeatedly perturbing the initial conditions of a single forecast model. Importantly, this approach provides robust estimates of forecast uncertainty, which supports human judgement and decision-making. Although weather forecasts and their uncertainty are also crucial for the weather-to-power conversion for RES forecasting in system operation, power trading and balancing, the industry has been reluctant to adopt ensemble methods and other new technologies that can help manage highly variable and uncertain power feed-ins, especially under extreme weather conditions.

In order to support the energy industry in the adaptation of uncertainty forecasts into their business practices, the IEA Wind Task 36 has started an initiative in collaboration with the Max Planck Institute for Human Development and Hans-Ertel Center for Weather Research to investigate the existing barriers in the industry to the adoption of such forecasts into decision processes. In the first part of the initiative, a forecast game was designed as a demonstration of a typical decision-making task in the power industry. The game was introduced in an IEA Wind Task 36 workshop and thereafter released to the public. When closed, it had been played by 120 participants. We will discuss the results of our first experience with the experiment and introduce some new features of the second generation of experiments as a continuation of the initiative. We will also discuss specific questions that emerged when we started and after analysing the experiments. Lastly we will discuss the trends we found and how we will fit these into the overall objective of the initiative which is to provide training tools to demonstrate the use and benefit of uncertainty forecasts by simulating decision scenarios with feedback and allowing people to learn from experience, rather than reading articles, how to use such forecasts.

1. Introduction

The challenges of climate change require unprecedented investments in renewable energy sources (RES) and a fundamental transformation of existing infrastructures such as the power grid. A key challenge is that RES like wind and solar power are characterised by an inherent inter-

mittency and variability that is unknown to fossil fuels and other weather-independent energy sources. To deal with the inherent variability of RES and seamlessly integrate them into the power grid requires a combination of weather forecasts and weather-to-power conversion methods. However, current approaches are based on deterministic weather forecasts, which provide no means to explicitly represent the varying uncertainty of forecasts that can make it difficult and costly to integrate high amounts of such resources into the electric grid.
As the penetration levels of RES increase and extreme weather conditions rise with the observed climate change, current methods have reached their limit, because they neither model nor convey forecast uncertainties. Instead, new methods for power production forecasts are called for that explicitly model uncertainty in order to make robust predictions and allow communicating forecast uncertainty to market participants and system operators [1]. This has been confirmed by Sweeney et al. [2] in their review on the *Future of Forecasting for Renewable Energy*, where they conclude that "..future forecast products will need to include probabilistic information, but deliver it in a way tailored to the end user and their specific decision-making problems.
Integrating uncertainty into the decision process is also important, because RES can be used at much lower costs than traditional energy sources due to their high flexibility - although only one-sided for down-regulation and with limited capacity for balancing power at the grid level.

Another aspect is that trusting in deterministic forecasts alone means that one ignores the underlying uncertainty. The world meteorological organisation (WMO) argues that "...if a forecaster issues a deterministic forecast the underlying uncertainty is still there, and the forecaster has to make the best guess at the likely outcome. Unless the forecaster fully understands the decision that the user is going to make based on the forecast, and the impact of different outcomes, the forecaster's best guess may not be well tuned to the real needs of the user" [3].

The WMO guide for communicating uncertainty also states that "...Uncertainty in the forecast can also arise from how the forecaster utilises the available information. Even if the model predictions are highly accurate, they must still be interpreted and translated by the forecaster into actual weather. This interpretation must then be rendered into a forecast, which in turn is received and interpreted by the user. Uncertainty can occur at each of these stages of the 'information chain'." [4]

2. Explaining the Need for the Initiative

The need to take into account and explicitly model forecast uncertainty is at the heart of many scientific and applied enterprises. For instance, the ever-increasing accuracy of weather forecasts has been driven by the development of *ensemble forecasts*, where a large number of forecasts are generated either by generating forecasts from different models or by repeatedly perturbing the initial conditions of a single forecast model. Importantly, this approach provides robust estimates of forecast uncertainty, which supports human judgment and decision-making (e.g.[5]). Although weather forecasts and their uncertainty are also crucial for the weather-to-power conversion for RES forecasting, the industry has been reluctant to adopt ensemble methods and other new technologies that can help manage highly variable and uncertain power feed-ins under different weather conditions. For instance, energy traders can use uncertainty forecasting to optimize the amount of power generation that they bid into the market, and transmission system operators can define the required reserve to account for the uncertainty in generation or be prepared for grid congestion much further in advance than with the currently used deterministic forecast. As a direct consequence of adopting these state-of-the art methods, the renewable energy sector forgoes critical potentials to reduce its vulnerabilities, build more robust prediction models, and improve judgment and decision-making of the involved parties.

There are two intertwined challenges responsible for the limited adoption of uncertainty forecasts in the renewable energy sector. First, current energy management systems (EMS) simply cannot operate with probabilistic forecasts. All processes in the EMS systems are designed as single value deterministic decision tools that are not capable of making simulation with ranges of possible outcomes or even different scenarios that reflect the uncertainty in the production of RES.

The second challenge is the human factor: New methods and approaches are bound to fail, if transmission system operators and other relevant agents do not know how to harness new forecast methods and systematically integrate uncertainty in their decision-making processes.

Probabilistic uncertainty forecasts have been shown to improve decisions in other weather related domains [6] and are likely to benefit power trading decisions as well. On the one hand, (a) probabilistic uncertainty forecasts contain information highly relevant to the decision [6]; on the other hand, (b) they make it easier for users to learn whether negative outcomes result from their decision strategy or the uncertainty of the forecasts compared to deterministic forecasts.

Our aim is to address these challenges in a unified and multi-disciplinary approach, bringing together hitherto separate fields (meteorology, behavioural insights, cognitive science, energy sector) and competencies. The overarching goal is to demonstrate the value of using ensemble forecasts in the RES sector.

3. The Games and Experiments

The objective of this initiative is to answer the most pressing questions that prevent the development and use of probabilistic forecast in the renewable energy sector by using behavioural decision experiments to simulate real-time problems for specific user groups using ensemble data and the corresponding power production forecasts.

The rationale behind the naming of "games & experiments" is that we design experiments in a game-like structure that is supposed to trigger some excitement in the participant's experience in the race for the highest scores or money earned by taking the best decisions, applying the best strategy or by just being lucky.

The results and feedback from these experiments can then be used to formulate strategies for further research and how to overcome the barriers that prevent the use of such forecast information and the harvest of the associated benefits from the additional information.

3.1. Background and Setup

The goal of the first experiment was to simulate a decision-making task in the power market, where there is high uncertainty of the production forecasts and where wrong decisions directly could be related to costs. Uncertainty in weather and also power production forecasts is not a new topic (e.g. [1], [7], [8], [9]) and has been discussed in many workshops of industry groups such as the Energy Systems Integration Group (ESIG) in the USA or the German Forecasting Platform (IFP) run by the Fraunhofer Institute of Energy economy and energy systems technology (IEE) and the German Weather Service (DWD) over many years.

Today, such information is readily available and there seem to be obvious advantages of using such information. However, it is also known that a number of factors influence human responses and decisions under uncertainty that go beyond a "rational" calculation of costs and probabilities. In order to investigate the factors that may underlie the hesitant reaction to the use of probabilistic forecasts observed in industry, we designed the first experiment in such a way that the participants had a direct comparison of deterministic and probabilistic information as input to their decision-making task.

3.2. First Experiment: Wind Power Trading for an Offshore Wind Farm

For the first experiment, we chose a trading situation that provided different types of information for the decision-making:

(i) deterministic forecasts of wind speed and wind power

(ii) probabilistic ensemble forecasts of wind speed and wind power

In the experiment, we assumed that most traders and balance responsible parties would agree that over time, the success of trading wind and solar power in a power market situation is related to the costs of the balancing power that is required to level out forecast errors. The income or cost profile is relatively strongly correlated with these balancing costs and that few events with high forecast error would be the driver for the costs and reduction of the income. We assumed that approximately 5% of the time, where the forecasts are off track account for 95% of the costs over a longer period of time, e.g. some months or a year. The assumption here is that it is more beneficial to reduce the balancing costs for large errors than seeking for a general forecast improvement of 1-2%.

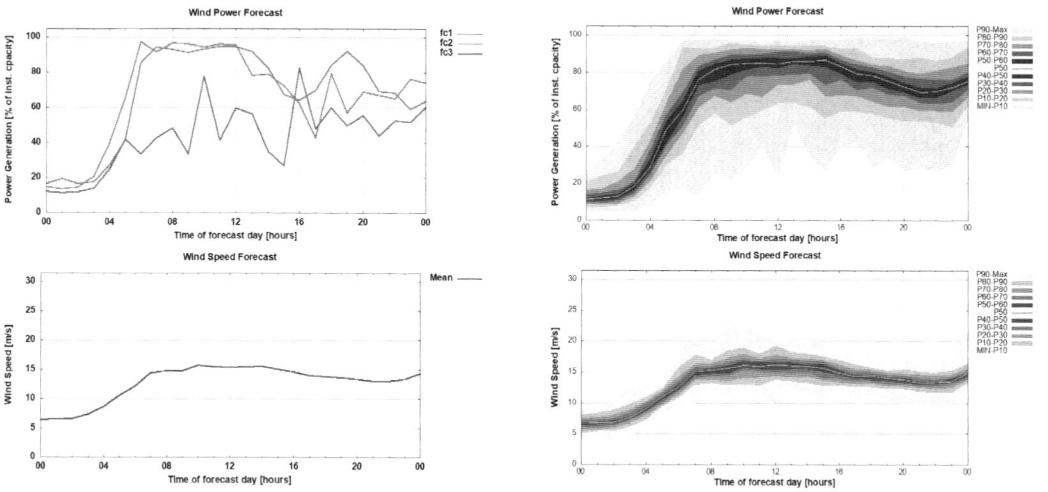

Figure 1. The deterministic forecasts (left) and the probabilistic forecast (right) as example of one of the chosen situations. For each type of forecast, the upper figure shows the wind power forecasts and the lower figure shows the wind speed forecast. The orange line in the wind speed forecasts marks the threshold (25 m/s) around which a high-speed shutdown can occur.

Fig. 1 shows one situation that was presented to the participants in the first example, where the deterministic forecast does not indicate or at least not clearly indicate the risk for a high-speed shutdown event, whereas the probabilistic forecast's upper percentiles P80-P90 and the maximum forecast in wind speed clearly indicated the risk, confirmed by the power forecast that shows the effect on the power generation with the percentiles P10-P40.

As shown in the example plot in Figure 1, we simplified the task by letting the users decide whether the generating power of an offshore wind park should be traded fully or partially given the possibility of a high-(wind)speed shutdown, where the wind park stops generating due to excessive wind conditions. Such so-called high-speed shutdown (HSSD) events occur in wind ranges between 21-27 m/s and are often referred to as cut-off wind threshold at 25 m/s. Because the wind turbines are not only calibrated to react on average wind speeds over a certain period of time, but also short-lasting wind gusts, there is considerable uncertainty on a wind farm to enter into high-speed shutdown (HSSD).

The cost function for the decisions to be taken in the experiment can be seen in Table 1. In the experiment, the participants had to decide, for each case, whether to trade 50% or 100%. The latter caused a penalty of 5.000 monetary units in case of the high-speed shutdown (HSSD) within the forecast period and a gain of 5.000 monetary units in case no HSSD occurred. When traded only 50%, the participants gained only 2.500 monetary units if no HSSD occurred, but did not lose any money, if a high-speed shutdown event occurred.

The rationale behind this cost-loss function was that in most applications, whether this is for trading purposes or balancing on system operation level, it is more expensive to buy balancing power than to have surplus generation. In the latter case, most units can be regulated down, which only causes a loss for not producing, but no extra cost. In the case of a lack of power in respect to the schedule, or bid in the market, due to a high-speed shutdown, the balancing power will be more expensive than the marginal costs for producing the power on the unit that shuts off. The simplification is therefore only on the timing, i.e. we do not request to ask for the specific hours in which the reduction takes place, but only whether any hour in the forecast time slot may experience a reduction. In other words, we look at the amplitude of the event, but ignore the phase.

All situations were real-world forecasts and reflect a decision that traders or operators are faced with in their daily work.

Trading amount	HSSD	No HSSD
100%	-5000	5000
50%	0	2500

Table 1. *Cost-Loss table for the "Offshore Wind Power Trading"experiment, where the participants are penalised when trading 100% and a high-speed shutdown (HSSD) event occurred and gain only 2500 monetary units, if they traded only 50% and no HSSD occurred.*

The first experiment showed us a number of interesting aspects, both from the participant's reaction and the results of the experiment. These lessons learned can be summarised as follows:

- A significant amount of the participants would like to have both probabilistic wind speeds and wind power generation forecasts or a deterministic "best guess" inside the uncertainty bands
- The improvement in decisions with additional probabilistic information was not equally strong in all cases, which points to the importance to understand the indicators in the forecast and the strategies that people use to make their decisions
- A situation with low visible uncertainty caused participants to expect no HSSD and to take a false risky decision, which raises the questions how threshold values are intuitively or rationally evaluated for uncertainty bands

3.3. Second Experiment - Wind Power Trading for a Wind Farm in complex Terrain

In the first experiment, we used an experimental design, where participants made decision based on deterministic forecasts first and in a second step were shown probabilistic ensemble forecasts. Each decision had to be confirmed or reverted based on the probabilistic ensemble forecast. With this setup we investigated, whether participants benefitted from the additional information of the probabilistic forecasts and whether the risk strategy changed.

One disadvantage of this design with updates was, that it does not allow to quantify whether or not participants will make better decisions by using uncertainty or probabilistic ensemble forecasts alone. Instead, the participants essentially decided whether to hold on to the previously made decision or change their mind. From a psychological perspective this is a critical aspect to consider when drawing conclusions from an experiment. For instance, often people do not update their beliefs sufficiently to change their minds. Thus, people may decided differently, independent of whether they were presented the probabilistic forecast or not.

For the second experiment, we instead used a design suited to evaluate the benefit of probabilistic forecasts independently (see Figure 2. The experiment has been setup in the following way:

(i) Every participant makes all decisions based on deterministic as well as on probabilistic forecasts. For each forecast type, the situations are presented in blocks, randomized among each other.
(ii) For every participant, the order of blocks is randomly chosen at the beginning.
(iii) For each decision the participants have to indicate how confident they were with the decision in a scale from 50% to 100%.
(iv) After each set of decisions, the participants are asked to describe the strategy and cues they have used in their decision-making.
(v) A unique ID allows participants to play multiple times with different nicknames in order to try out different decision strategies.

As shown in the schematic graph in Figure 2, the forecast situations will be randomized in both runs so that participants are not getting the same sequence of forecasts to prevent that they remember the situations. In both runs, participants only take their decisions on the basis of either deterministic or probabilistic forecasts. The goal is to be able to identify the benefits, if any, by using the probabilistic forecasts and deterministic forecasts on their own and explore potential problems in the use of the probabilistic forecasts in more detail. The participants are additionally being asked after each decision how confident they are with the decision they made. With this, we want to investigate, whether there is a difference in confidence level between decisions taken on the basis of deterministic versus probabilistic forecasts. Another aspect that is going to be investigated is the way participants cope with failures or false alarms; that is, how participants react after taking a wrong decision and whether there is a difference when being presented deterministic or probabilistic forecasts. The use of forecasts from a real-time environment enables us to choose forecasts that seem to have clear indicators for a specific extreme event, but did not turn out to be one such event and vice versa. These so-called false alarm events and the reaction to such false alarms by the decision-makers are often highly underestimated in real-time environments to have impact on decision-making. The psychological effect of a false alarm can also turn out to be very different from person to person. By getting a better understanding of such reactions, we can find strategies how to adapt forecast presentations and train decision-makers, but also their managers and the forecast providers to better understand the difficulties coming along with such situations.

When decisions are made for all 40 cases, participants are asked a number of general questions; first they are asked about the cues or strategies that they used when making their decisions and how they found that this worked. Thereafter, the confidence level is requested once more, i.e. which type of forecasts the participants found useful in terms of deterministic forecasts and probabilistic forecasts and which of these forecasts they would want to have or consider irrelevant.

Lastly, the participants are asked about their background and the experience in the sense of time working in an area that is related to such decision-making, the age of the person and the gender. In that way, we can study various psychological aspects in the cognitive science area that may provide clues on how to best present the uncertainty of forecasts for decision-makers to benefit and to keep focus and confidence in their roles without becoming over-confident.

Journal of Physics: Conference Series **2151** (2022) 012014 doi:10.1088/1742-6596/2151/1/012014

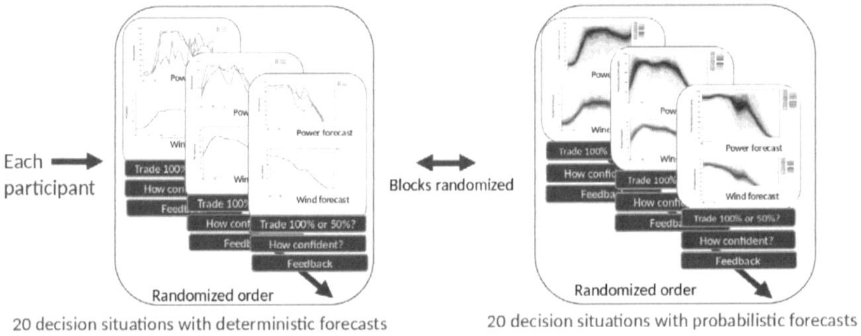

Figure 2. Setup of the second experiment, where the cases are separated between decisions made on the basis of deterministic forecasts (left side) and decisions made on the basis of probabilistic ensemble forecasts (right side).

4. Lessons learned and next Steps

The first experiment in the series has revealed a number of interesting aspects regarding decision-making with and without uncertainty information from probabilistic ensemble forecasts. Although we can conclude that there is a potential benefit, and most participants also confirmed in the follow-up questionnaires that they clearly prefer to have probabilistic forecasts at hand when making decisions on extreme events, there are still many open questions. Some of these will be answered in the second experiment, for instance, the impact on decisions, when presented by either probabilistic or deterministic forecasts on its own.

Nevertheless, when we closed the experiment, where 120 participants had gone through the 20 decision cases, there was a clear difference in the final balance scores and the distribution of these scores. Figure 3 shows the result of that final balance as whisker box plots and histogram of the income. The distance between the lower and upper limit of the boxes shows the interquartile range (IQR) of the distribution (distance between the 25th and 75th percentiles), where the median is represented by the horizontal line. The upper (lower) whisker extends from the box to the highest (lowest) value within 1.5 * IQR. It can be seen in Figure 3, that the interquartile range for the probabilistic ensemble forecasts was with 33% significantly smaller than for the decisions based on deterministic forecasts. Also, the amount of participants that reached the highest scores, which corresponds to the generation of the highest income from the traded cases according to the applied cost function in Table 1, have all reached them when using the probabilistic ensemble forecasts for their decision.

The most obvious conclusion from the higher income generated with probabilistic forecasts suggests that there exists a benefit of using probabilistic ensemble forecasts for this type of application, i.e. trading, where such extreme weather events are responsible for the typical 5% of the largest error, leading to large balancing costs; it also indicates, seen in the smaller spread, that people tend to make more similar decisions when presented with probabilistic ensemble forecasts in their decision-making process. For many applications, where the final decisions are made by human decision-makers, this is an important aspect to consider, especially, if decisions are supposed to be fair, transparent or even just consistent.

One very interesting aspect of the latter observation is the phenomena of personal bias, well known from e.g. environmental assessment (e.g. [10]) or insurance case assessments (e.g. [11]. In other words, if the spread of the final balance decreases when using uncertainty forecasts, it would be a benefit for any organisation working with a group of decision-makers that work e.g. in shifts. Reduction of the so-called personal bias would be a benefit that goes beyond financial income. Once we have a large enough sample of participants in the second experience,

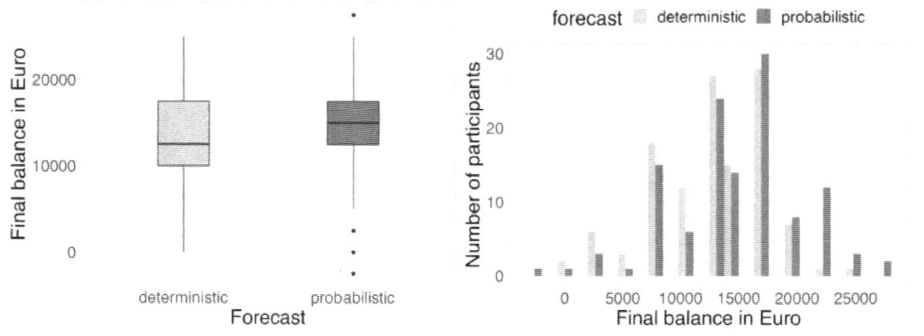

Figure 3. Final Balance of the first experiment, when closed, played by 120 participants; shown as whisker plot (left) with interquartile range, showing the result of 50% of the participants and as histogram (right).

this aspect will also be compared to the first experiment. Here, it will be interesting to explore the difference related to the decision-making process, i.e. when making all decisions based on deterministic or probabilistic forecasts. Especially in a transition phase, where the industry starts to use uncertainty forecasts, it could well be, that the benefit is stronger, if participants can change their mind when presented with uncertainty forecasts, i.e. reduce failure of the risky decisions.

4.1. Additional Forecast Information for the Decision-making Process

The first two experiments have been solely focusing on decision-making with the help of graphical time series forecasts for both, the deterministic and the probabilistic forecast types. However, we have not been investigating, as described e.g. in [12], whether the type of representation of the uncertainty also could have an impact on decisions, i.e. whether some or all end-users may be more likely to make better decisions, if they would get the probabilistic information additionally in a text or table format.

Or, if for example a graph or a series of graphs of the large-scale weather situation in a horizontal plot as shown in Fig. 4 could be useful for the decision-making process. The example shows the wind speed (colours), wind direction (arrows) and isobars, joining together places of equal atmospheric pressure, for the time of the highest probability of a high wind speed shutdown event from the example in Fig. 1 as mean (large left graph) of 75 forecasts and the minimum (upper right) and maximum (lower right) thereof, respectively.

5. Conclusion and Future Work

The results and reaction of participants from our first two experiments point towards an industry that has understood the need for uncertainty forecasting in the handling of renewable energy sources (RES). Nevertheless, the adaptation of uncertainty forecasts in the daily practices is still somewhat slow. One reason for this is the lack of knowledge about the different sources of uncertainty forecast products and their specific applicability, even though information is available in literature e.g. [1], [13]. The translation from research literature to industry guidelines together with the increasing need for uncertainty estimates in the power system operation may be necessary for the industry to make the leapfrog step into the probabilistic design of the underlying, e.g. EMS, software. The IEA Wind Task 36 has already started to incorporate uncertainty forecasts and currently applied probabilistic forecast in the recommended practice

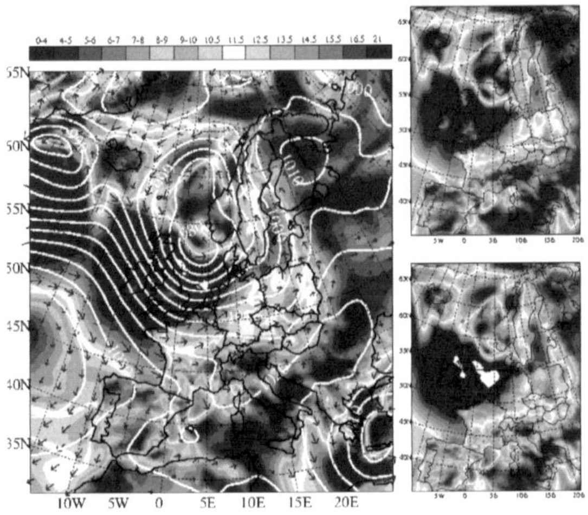

Figure 4. Weather maps of wind speed (colour scale), wind direction (arrows) and isobars. The large graph is the EPS Mean of 75 members, the upper right graph is the minimum and the lower right graph the maximum wind speed of the 75 members in each grid point.

guideline for forecast solution selection [14] in it's updated version that is currently underway to support the industry in that process.

In our future work, we can with some certainties assume a general acceptance and request for probabilistic forecasts in energy related decision-making. Therefore, we will in the future focus more on low probability, but high-impact scenarios and which tools are most effective for the decision-making on the basis of probabilistic forecasts for typical decision problems. The following is a list of topics that we want to follow in the next period of the task 36:

(i) Provision of other graphics, e.g. weather maps of large-scale weather or presenting percentiles versus all ensembles members to the decision-maker to investigate how decisions may change when more or less ensembles members are presented

(ii) Provision of probabilities in form of text or combinations of graphics and text

(iii) Study of false alarm situations, where the reaction of participants is specifically analysed in conjunction with the correctness of the decision

(iv) Investigate whether decision-makers are more risk averse or prone given probabilistic forecasts and whether the amount of uncertainty makes a difference

(v) Scenarios with extreme events, where different types of uncertainty forecasts are used and where some methods are not able to capture the event

(vi) Scenarios with extreme balancing power prices (this sometimes occurs in the Nord Pool or EPEX market), where forecast errors in one direction can be highly penalized

Cases i and ii would be interesting to study, which tools are most useful for decision-making. Here, participants could for example be presented with different types of uncertainty forecasts, all ensemble members versus aggregated percentiles, as well as additional graphical or textual interpretations of the same event to make a decision.

Case iii and iv is an interesting investigation (1) when decision-making takes place on the basis of small or large probabilities and (2) how decision makers react, when they have taken a wrong decision. Part of the latter are we already investigating in the second experiment. What is

missing in that experiment is the discrete knowledge of the probability or a prior taken decision of the probability that leads to a certain event and the consequences thereof.
Case v are scenarios, where we investigate how difficult it is to evaluate, whether or not a certain methodology is suitable for an application with extremes with e.g. long return periods and where e.g. statistical methods fail, because (1) the event was never observed in historical data or (2) the temporal dependency structure is not modelled by the statistical approach used. Such an experiment would allow investigating how well different modelling approaches are understood and applied correctly when introduced to the industry. Since the science behind the probabilistic approaches are rather complicated, the IEA Wind Task 36 has started to establish documentation about different approaches broken down to a level that addresses the typical educated power engineers. However, it is unclear whether this type of scenario and its hidden uncertainty play a role in the currently observed barriers of implementing probabilistic applications. *Case vi* is more directed to specific user groups and applications. A good possibility could be to use the probabilistic game in a utility theory setup to study the choice between a feed-in tariff y (guaranteed income) or direct participation in the electricity market with probability p of earning less than y or $1 - p$ for winning more than y.

Acknowledgment

The authors would like to thank all the participants of the experiment and the IEA Wind Task 36 group for valuable input in the preparation of the experiments. Corinna Möhrlen and Gregor Giebel were supported by the Danish EUDP project IEA Wind Task 36 Phase II Danish Consortium under contract no. 64018-0515. Nadine Fleischhut was supported by the Hans-Ertel-Centre for Weather Research of the German National Weather Service.

References

[1] Bessa, R. J., Möhrlen, C., Fundel, V., Siefert, M., Browell, J., E.Gaidi, S.H., Hodge, B.-M., Cali, U., Kariniotakis, G. (2017), Towards improved understanding of the applicability of uncertainty forecasts in theelectric power industry, Energies, vol. 10, no. 9, p. 1402.
[2] Sweeney, C, Bessa, RJ, Browell, J, Pinson, P. (2020) The future of forecasting for renewable energy. WIREs Energy Environ., no. 9:e365, https://doi.org/10.1002/wene.365
[3] Gill, J., Rubiera, J. Martin, C. Cacic, I., Mylne, K., Dehui, C., Jiafeng G., Xu, T., Yamaguchi M., Foamouhoue, A. K., Poolman E., Guiney, J. (2008), Guidelines on communicating forecast uncertainty," World Meteoro-logical Organization, Tech. Rep. WMO/TD No. 4122.
[4] World Meteorological Organization (WMO), (2012), Guidelines on Ensemble Prediction Systems and Forecasting, Tech. Rep. WMO-No. 1091.
[5] Pahlow, M., Möhrlen, C., Jørgensen, J.U. (2008), Application of costfunctions for large-scale integration of wind power using a multi-scheme ensemble prediction technique, in Optimization Advances inElectric Power Systems, E. D. Castronuovo, Ed.ISBN: 978-1-60692-613-0: NOVA Publisher NY, ch. 7, pp. 151–180.
[6] Joslyn S, LeClerc J. (2013) Decisions With Uncertainty: The Glass Half Full. Current Directions in Psychological Science, 22(4):308-315, doi:10.1177/0963721413481473
[7] Fundel, V.J., Fleischhut, N., Herzog, S.M., Göber, M. and R. Hagedorn, (2019), Promoting the use of probabilistic weather forecasts through adialogue between scientists, developers and end-users, Quarterly Journal of the Royal Meteorological Society, vol. 145, no. S1, pp.210–231.
[8] Murphy, A. H. and Ye, Q. (1990), Optimal decision making and the valueof information in a time-dependent version of the cost-loss ratiosituation, Monthly Weather Review, vol. 118, no. 4, pp. 939–949.
[9] Möhrlen, C., Pahlow, M., and J. U. Jørgensen, (2012) Untersuchung verschiedener Handelsstrategien für Wind- und Solarenergie unter Berücksichtigung der EEG 2012 Novellierung, Zeitschrift f. Energiewirtschaft, vol. 36, no. 1, pp. 9–25.
[10] Glenn W. Suter II and Susan M. Cormier (2015), Bias in the developmentof health and ecological assessments and potential solutions, Human and Ecological RiskAssessment: An International Journal, DOI: 10.1080/10807039.2015.1056062
[11] Pinquet, J., Ayuso, M., and Guillén, M. (2007). Selection Bias and Auditing Policies for Insurance Claims. The Journal of Risk and Insurance, 74(2), 425–440. http://www.jstor.org/stable/4138443

[12] LeClerc, J. and Joslyn, S. (2015), The Cry Wolf Effect and Weather-Related Decision Making. Risk Analysis, 35: 385-395. doi:10.1111/risa.12336.

[13] Haupt, S.E., and Garcia Casado, M. and Davidson, M. and Dobschinski, J. and Du, P. and Lange, M. and Miller, T. and Mohrlen, C. and Motley, A. and Pestana, R. and Zack, J., (2019), The Use of Probabilistic Forecasts: Applying Them in Theory and Practice, IEEE Power and Energy Magazine, vol. 17, no. 6, pp. 46-57, Nov.-Dec. 2019, doi: 10.1109/MPE.2019.2932639.

[14] Möhrlen C. and Zack J.W. and Giebel G. (2019) IEA Wind Task 36 Recommended Practice on Renewable Energy Forecast Solution Selection, IEA Wind Task 36, International Energy Agency Technical Cooperation Programm Wind, https://iea-wind.org/task-36/task-36-publications/recommended-practice/,

AUTHOR INDEX

Institute of Physics
Dirac House, Temple Back
Bristol BS1 6BE UK

ISSN: 1742-6588
ISBN 978-1-7138-5122-6
9 781713 851226

CURRAN ASSOCIATES INC.
proceedings.com

Sixteenth International Conference on Advances in Computer-Human Interactions (ACHI 2023)

Held at DigitalWorld 2023

Venice, Italy
24-28 April 2023

IARIA